Burke Museum of Natural History and Culture Research Report No. 8

Vashon Island Archaeology
A View from Burton Acres Shell Midden

Edited by Julie K. Stein and Laura S. Phillips

Burke Museum of Natural History and Culture
Research Report No. 8

Burke Museum
Seattle, Washington

Distributed by the University of Washington Press, Seattle and London

Library of Congress Cataloging-in-Publication Data
Vashon Island archaeology: a view from Burton Acres Shell Midden/edited by Julie K.
Stein and Laura S. Phillips.
     p. cm. -- (Burke Museum of Natural History and Culture research report; no. 8)
  ISBN 0-295-98287-X (alk.paper)
  1. Puyallup Indians--Antiquities. 2. Excavations (Archaeology)--Washington
(State)--Vashon Island. 3. Vashon Island (Wash.)--Antiquities. 4. Burton Acres Shell
Midden (Wash.) I. Stein, Julie K. II. Phillips, Laura S. III. Series.

E99.P8 V37 2002
979.7'77--dc21

2002026068

The paper used in this publication meets the minimum requirements of American
National Standard for Information Sciences — Permanence of Paper for Printed
Library Materials, ANSI Z39.48-1984.

King County
Landmarks & Heritage Commission
Hotel/Motel Tax Fund

Burke Museum Publications Program

Dr. George MacDonald, Director
Dr. Susan D. Libonati-Barnes, Editor

Front Cover: *"Arched Salmon"* (Shaun Peterson, 2002, member of Puyallup Tribe of Indians)

Contents

Appendices   Available under separate cover: Burke Museum of Natural History and Culture
Department of Archaeology
Box 353010
University of Washington
Seattle, WA 98195-3010
recept@u.washington.edu

Contributors

Timothy Allen
Ohio Historic Preservation Office
567 E. Hudson St.
Columbus, OH 43211-1030

Kristine Bovy
Department of Anthropology
Box 353100
University of Washington
Seattle, WA 98195-3100

Virginia Butler
Department of Anthropology
P.O. Box 751
Portland State University
Portland, OR 97207

LLyn De Danaan
LLyn De Danaan, Ph.D. and Associates
S.E. 142 Oyster Beach Rd.
Kamilche, WA 98584

MaryAnn Emery
Department of Intercollegiate Athletics
Box 354070
University of Washington
Seattle, WA 98195

Robert Kopperl
Department of Anthropology
Box 353100
University of Washington
Seattle, WA 98195-3100

Mary Parr
Pastime Software
17104 West 65th Circle
Arvada, CO 90007

Laura S. Phillips
Burke Museum of Natural History and Culture
Box 353010
University of Washington
Seattle, WA 98195

Julie K. Stein
Department of Anthropology
Box 353100
University of Washington
Seattle, WA 98195-3100

Nancy A. Stenholm
Botana Labs
12329 26th Ave. N.E.
Seattle, WA 98125

Judy Wright
Puyallup Tribe of Indians
2002 E. 28th St.
Tacoma, WA 98040

Preface

Today, archaeologists work alongside tribes to protect the archaeological resources of our region. We often have different opinions about how best to protect archaeological sites. Only by working together can we address our many concerns and arrive at the best ways to care for this state's cultural heritage. This publication attempts to continue these cooperative efforts. The first two chapters describe the Vashon Island Archaeological Project and its goals. Chapter 1 introduces the perspective of Judy Wright, a tribal member of the Puyallup Tribe of Indians and member of the project's advisory committee, who was instrumental in permitting us to do the project and provided us with guidance along the way. Chapter 2 was written from the perspective of Julie Stein, an archaeologist and the principal investigator for the project. It is essential to the understanding of archaeology in our region that multiple perspectives be heard.

Archaeologists have done a poor job of educating the public about the need to preserve the cultural heritage of our region. Too often, archaeological sites are destroyed out of ignorance. Construction workers and private property owners often are reticent to report sites, because they think their projects will be halted. In truth, a few projects *are* temporarily stopped so that sites can be investigated before they are destroyed. More often, however, projects are re-routed to avoid sites, or construction work is redirected elsewhere within the project area, so that employment is not interrupted.

Archaeologists, working with tribes, government agencies, and project managers, protect archaeological resources most effectively when sites are reported and investigated before construction projects occur. Few of the negotiations regarding site stewardship and investigation occur with public knowledge. This is necessary, in part, to expedite construction, as well as to protect sensitive tribal issues, particularly when human remains are encountered.

The public is provided very little knowledge of cultural resources. Ultimately, the preservation of our heritage is at stake, and local communities need to be entrusted with the stewardship of these sites.

Archaeological sites are ubiquitous in Washington. The locations of these sites are not discolosed to the public, but records are kept at the Office of Archaeology and Historic Preservation in Olympia, with the intent that artifact collectors cannot destroy this cultural heritage. Federal, state, and local governments have access to this information, in order to ensure that construction of buildings, roads, and other changes to the landscape do not inadvertently destroy these archaeological sites.

The management of cultural resources (which includes historic properties, as well as archaeological sites and traditional cultural properties) affects the history and heritage of the peoples of Washington, yet very few citizens are even aware of the richness of these archaeological resources.

The intent of archaeologists, tribes, and government agencies to protect sites by restricting access to site locations is admirable and seems necessary. Still, despite our concerted efforts at protection, looting continues with disturbing regularity. In fact, people who illegally collect artifacts are often just as good as archaeologists at finding sites. Recently, a burial in King County was destroyed by two backhoe operators, who told neighbors that they were testing the soil, when they were actually illegally digging up the burials and funerary objects. Had the neighbors known about the site, they could have prevented its desecration.

In truth, however, this is a double-edged sword. If the public is provided with site location information, then a few bad people may destroy the sites. However, since a few bad people already know where most of the sites are, perhaps providing this information to everyone, along with education about site stewardship, will ultimately provide the public with the tools to protect the sites.

At the Burton Acres Shell Midden, we took this second approach. By conducting a public education project, and asking the community for support, we hoped that the Vashon Island community would recognize their stewardship role in the on-going protection of the site. During the excavation, we had the unique opportunity to be able to devote one-on-one time with individual members of the public. We answered a full range of archaeology questions, ranging from questions about Egyptian and Chinese archaeology to questions about Washington State laws that pertain to archaeological sites. In addition, we distrib-

uted information flyers, brochures, and articles about archaeology and site preservation in Washington.

The excavation provided a venue for educating the public about the stewardship of the Burton Acres Shell Midden. It was also meant to alert the people of King County about the destruction of sites on their own property. The many privately-owned archaeological sites in King County are disappearing rapidly due to ignorance and irresponsible development, as well as natural causes. There are many people on Vashon-Maury Island who have been well aware that they have an archaeological site on their property, yet they had never been given any information about the site or how best to protect and preserve it. Few if any of the people who visited Burton Acres during the archaeological investigations knew that King County offers property tax incentives for land owners who have sites located on their property (Department of Development and Environmental Services Current Use Assessment – Open Space Fact Sheet).

By the end of the project, we knew we had succeeded in our education goals. Numerous people took part in an effort to record sites on their property. And, Vashon residents continue to call us with concerns about site erosion and potential vandalism at the Burton Acres Shell Midden. The people of Vashon-Maury Island understand that stewardship is a rewarding process, and that we cannot afford to lose the cultural heritage of Puget Sound.

Acknowledgments

We owe the success of the excavations at the Burton Acres Shell Midden to the collaborative efforts of the Thomas Burke Memorial Washington State Museum, King County Landmarks and Heritage Commission, the Puyallup Tribe of Indians, Vashon Park District, and McMurray Middle School. These entities were instrumental in helping us to obtain permission from Washington State to investigate the site. But, more importantly, they supported and encouraged us to bring archaeology to the public.

Julie Koler and Holly Taylor of King County's Historic Preservation Program first approached us to conduct a site investigation in 1996 to determine if the site had enough integrity to be eligible for nomination to Landmark status. When Julie Stein suggested that we invite the public to participate in the project, Koler and Taylor were thrilled, and immediately began performing miracles to obtain funding and community support to bring the project to fruition. They are committed to the protection and preservation of archaeological sites, and have continued to support us through the entire project. Their support includes the publication and dissemination of this book, which was funded by a grant from the King County Landmarks and Heritage Commission, 2000 Special Projects Program.

The Puyallup Tribe of Indians has also been supportive from the moment we contacted them about the project. Three individuals, in particular, worked hard to ensure that this project was done in a way that would benefit all of the people of Puget Sound. Carol Ann Hawks, Marguerite Edwards, and Judy Wright were willing to share their concerns about our project, and to educate us about the ancestors of the Puyallup who lived on Vashon-Maury Island. We thank them for their commitment to the project, and for their trust in us. Judy Wright's willingness to author Chapter 1 of this publication was especially important to us, as well as to the future of cooperation between archaeologists and tribes.

Roxanne Thayer, a teacher at the McMurray Middle School, was working on her own project – a video documentary called *Vashon's and Maury Islands: Hands Across Time* – at the same time that we were embarking on ours. She quickly became an essential part of the Burton Acres project, asking her students to join our excavation as tour guides and assistants to the archaeologists. Thayer has an unending amount of energy and a gift for imparting that enthusiasm to her students. She and her students were a wonderful asset to the excavations. But, her advocacy for our project didn't stop there. She and Michael Kirk, principal of McMurray Middle School, also provided the day-to-day essentials: a place to live. We are indebted to them for providing us with their warm, comfortable school where we could not only sleep and eat, but where we could set up a lab and storage space.

The Vashon Park District was crucial to our project, and we thank them for granting us permission to investigate the Burton Acres Shell Midden.

Several organizations on Vashon-Maury Island helped to fund our project, and provided us with historical information about the island. These include the Vashon-Maury Island Heritage Association, Vashon Friends of the Library, and the Vashon Island Rotary Club.

We thought one of the hardest parts of this project was going to be asking the community for help to fund the project. We were wrong. We could hardly believe that so many people were willing to help us. Several groups and individuals were so committed to the project that they provided us a space to raise funds. Our dear friends Deehan Marie Wyman and Mary Dunnam not only provided fundraising space, but the people and moral support. Vashon Allied Arts and the Vashon Kiwanis Club also welcomed us into their space, and helped us to gain the support of Vashon-Maury Island residents.

Their enthusiasm and support was infectious, and we can't thank them enough. Many others eagerly provided support. There are too many involved to name them all, but the list below includes those that contributed significant help towards the project.

Don and Jane Abel; Nancy and Craig Abramson; Jean and Ray Auel; Mary Jo and C.W. Barrentine; Barbara and Jonathan Bayley; Charlotte Bennett; G.M. Boeing; Tom and Mikki Brain; David and Mary Anne Buerge; Cherry and James Champagne; J. David Cole; Taffy Crockett; Mary and James Dunnam; Donald and Patricia Eastly; Ed Palmer Construction, Inc.; Richard S. Erwin; George Evanoff, Jr. and Maria Evanoff; Ellen Ferguson; Joy A. Goldstein; Arnold and Iola Groth; Cheryl E. Gudger; Teri-Ann R. Guthrie; Jennie A Hodgson; Rayna and Jay Holtz; Fred and Jeanne Howard; Hugh and Jane Ferguson Foundation; Darlene and Jerry Kenney; Michael and Patricia Kirk; Karen Knutzen; Sheldon and Susan Kopperl; Alice Larson; Marion and Bonnie Larson; The Law Offices of Fiore J. Pignataro, Inc., P.S.; Lizz Maunz; Wallace and Carolyn May; Lee A. Miller; Mark R. Musick; Louise Olsen; Pacific Research Laboratories, Inc.; John and Georgia Ratzenberger; Robin Home and Terry Friedlander; Anne Romano Sarewitz; George and Lauren Schuchart; Smith & Koch, P.S.; Stephen and Lucinda Stockett; David and Anna Swain; Vashon Friends of the Library; Vashon Island Rotary Club; Vashon Kiwanis Club; Vashon-Maury Island Heritage Association; Edward J. Wachter; Shirley J. Wetzstein; John and Ina Whitlock; Andrew Williams; Edith Williams; Mary and C.J. Worm; Wyman Youth Trust; Deehan Marie Wyman; Jean L. Young; Institute for Ethnic Studies in the US.

We also thank the authors of this publication for their contributions and dedication to the project. Their analyses are the results of many long hours in the lab and at the computer. Their work, however, could not have been completed without the help of many other experts. Many thanks to all of them. Stewart Pickford, UW Professor at the College of Forest Resources, willingly and enthusiastically volunteered his time and equipment, to help us survey and map the site. Tim Allen, UW archaeology graduate student, took this data and created an invaluable map of the site. Lance Lundquist and Elizabeth Martinson, University of New Mexico graduate students, drew the stratigraphic profiles in record time. Lundquist also spent hours magically creating a database for us. Arn Slettebak, Burke Museum Curator of Exhibitions, readily identified the pin-fire cartridge. Edward Bakewell, UW archaeology graduate student, provided identifications of the lithic material types. John Rozdilsky, Burke Museum Mammalogy Curatorial Assistant, allowed us to use the Burke Museum mammal collections for comparative analysis. Donald Grayson, UW Professor of Archaeology, and Michael Etnier, UW archaeology graduate student, were always on hand to identify mammal bones for us. Virginia Butler, Portland State University Professor of Archaeology, worked extensively with Robert Kopperl, UW archaeology graduate student. We cannot thank her enough for providing expert advice and instilling confidence. Jose Orensanz, UW Fisheries Research Scientist, provided expert identifications of shellfish remains for our comparative collection. Ronald Eng, Burke Museum Geology Collections Manager, offered the use of the malacology collection and library. We thank Ray Pfortner for his photographic documentation of the project (27520 94th Ave. S.W., Vashon Island, WA 98070). Chris Lockwood did the layout of the book and made it beautiful. Catherine Foster finished the job with attention to details. Betsy Blinks and Kate Gallagher checked the text for consistency. Special thanks goes to Chris Lockwood and Kate Gallagher who worked tirelessly for two years editing the text and graphics, and kept this project moving.

Thanks, especially, to Mary Parr for jumping on the plane from Hawaii to help us on this project. We could not have done it without her.

Finally, this archaeological project could not have been undertaken without the help and support of the people of Vashon-Maury Island. We hope this project has inspired them to protect the archaeology of the islands for the future.

Vashon Island Archaeology

1

The Project:
A Tribal Perspective

Judy Wright

From the inception of this project, the Puyallup Tribe and archaeologists worked together. Even though we had different opinions, each was free to express their own view, and each has learned from the other. The tribal perspective is important for an accurate understanding of the past and begins our report of the Burton Acres Shell Midden

When we were first approached with the idea of the Burton Acres Archaeology project at Vashon Island, we met with representatives from both the Burke Museum and the King County Historic Preservation Office. For us, it seemed as if the project had both a plus side and a negative.

Historically, we have known of the presence of our ancestors on the island, and the significance of proving that has been extremely important to us. Our records verified the existence of the rich culture of our people at Vashon.

Our legends and mythology also substantiated our history on the island. There are also present living members of our tribe who trace their heritage to the places at Vashon they have known historically as home. Oral tradition of our presence on the island has been passed down by many families.

Most Indian tribes find it extremely hard even to participate in the consideration of such projects, and consenting to it was very difficult. We, as Indian people,

have the deepest respect for the places that were occupied by our ancestors; we consider these sacred sites. Historically, relations between the scientific community of anthropology and archaeology and our people have been tenuous at best. Our hesitation to work within this structure through time has not necessarily been to our benefit. Educators in these fields do not necessarily agree and have the same philosophy as to the very essence of our being. Our creation stories, legends, and mythology — the history of our people handed down to us by our people — place us here from the beginning. We do not believe that we migrated here from Africa, Siberia, Asia, or any other place.

Finally, after years of persistent theories of a land bridge, the scientific community is beginning to look at alternatives regarding the prehistory of America and the evolution of the people who trace their being to this country. Information is beginning to emerge that challenges long held theories of the peopling of America, which may have occurred much earlier than

Chapter opening photo: Mary Sportsman, Judy Wright, and Lina Landry arrive at the Burton Acres Shell Midden to participate in the excavation.

Figure 1.1 Members of the Puyallup Tribe of Indians bring clams and salmon to celebrate the collaboration between tribal members, University of Washington archaeologists, King County preservation staff, McMurray Middle School students, and residents of Vashon Island.

previously projected. Still other data are suggesting that sites significant to this debate may have been inhabited thousands of years earlier than initial hypotheses have proposed. Several sites are being revisited: one, at Monte Verde in Chile, and another, at Meadowcroft Rockshelter, near Avella, Pennsylvania. With absolute recognition from our standpoint, that we know very little about archaeology, it is with great pleasure that we hear the dogmas previously professed are being carefully re-examined.

Concerns have emerged regarding archaeologists who support the developers' agenda of hasty construction. There are instances where it appears that in that rush, valuable data are lost, compromised, and disregarded. As observers of these practices, and knowing the importance of exhausting all avenues, it has been hard to have healthy relationships with common goals. We, too, have a thirst to know, and to prove our theories. We know that our own acceptance of our beliefs will not necessarily be taken at face value. However, our stories also have merit and deserve consideration. Many known sites have previously been damaged or destroyed.

The forever imminent danger of an archaeological project that possibly could result in disturbing human remains of our long departed is so undignified and distasteful to us as Indian people, it weighed heavily upon our minds. Our culture has taught us reverence for these places, and our traditions have taught us that it would be disgraceful to intentionally disturb the resting places of our people. The historic properties are plentiful. We have had the misfortune of living in close proximity to a community that has forged ahead blindly and had little or no regard for the presence of these sites, and the rich culture of our people.

Several issues also surfaced that needed to be taken into consideration. Erosion that was occurring, and the imminent danger that the site could in fact disappear. We recognized the potential benefit in allowing the professional community to proceed with this project. In fact, we decided we could not risk that the site not be analyzed by the professional community because of the potential of erosion at the site.

The Puyallup Tribe has been in the position of always having to prove that our people have known these areas as their homelands. We have had to gather evidence to prove we were fishermen, clam diggers, hunters, traders, and gatherers. Constant pressure prevails for our Tribe to prove our very existence. Being somewhat familiar with both the written record and the testimony of our elders, it seemed to us that perhaps through this project we might bridge some unfortunate beliefs held by others, that our people only visited places such as Vashon Island, when we know that it was our ancestors' home since time unknown.

We were perhaps one of the first tribes to work out a Memorandum of Agreement based on direct negotiation about areas of concern, and not based on the intervention of a federal government agency. This agreement allowed the tribe to endorse the project. We needed the evidence to document our presence on the island.

It is our belief that our people's feet touched the shores, breathed the air, felt the peace and serenity of silence, witnessed our beloved Mt. Takoma, and were thankful to the Creator for providing the abundance of nature. These were legendary times of happiness and fruitful living. The people took care of one another, remembered the long departed souls of their relatives, and cherished the land and the burial places of those gone to the other side. With this in mind, it was difficult for us to participate in the exploitation of sacred sites; it is comparable to sacrilege, as others would view desecration of their holy altars. Many do not understand this, and it has created a great deal of misunderstanding. We, as Indian people, seem to be constantly on the defensive in our own country, being asked to prove to others, that we were for example, clam diggers, fishermen, etc. This has created financial hardships for our tribe and consequently also the people. It has necessitated going to court, hiring lawyers, anthropologists, and experts with credentials to prove and validate what we already know.

It makes little difference that the old homes and historic properties no longer seem to exist for many. They can be called by different names, and the ownership seemingly belongs to others. We will remember where they were, and still claim them as our homes. They are the usual and accustomed sites of our ancestors.

Were it not for the sensitive nature of Dr. Julie Stein, Archaeologist from the Burke Museum, a professional who earned our respect and trust, we most likely could not have endorsed the project. She is an educator with high standards who saw the project had merit both as an aid in teaching the public and in raising the level of conscious awareness in so many. Her conduct has allowed us to have some semblance of trust, where previously there was a very thin thread. Her staff and volunteers conducted themselves in the most respectful and responsible manner. The project has been a success, and we personally thank her for all that she has done to make it so. She has been a credit to her colleagues.

Holly Taylor and Laura Phillips were also instrumental in seeing the project through. They made a very difficult job immensely easier. Both are to be commended, in our opinion, for a job well done. Roxanne Thayer worked tirelessly to see her project to completion, and it will assist in opening the door for an honest effort at looking into the past from a very different perspective.

We would like to thank King County Historic Preservation Officer, Julie Koler, for the recognition of the Puyallup Tribe and our Government from the outset. All of the participants who worked diligently on this project made it an enjoyable time, and we feel it was successful (Figure 1.1).

Photo courtesy of Ray Pfortner.

2

The Project: An Archaeological Perspective

Julie K. Stein

With the assistance of members of the Puyallup Tribe of Indians, University of Washington archaeologists and students from Vashon/Maury Island schools, a set of techniques were developed to educate the public about archaeology. The techniques follow the methods of professional archaeologists and permit people to understand the entire archaeological investigation from excavation to curation.

The Burton Acres Shell Midden (45KI437) is located at the easternmost tip of the Burton Peninsula, within Quartermaster Harbor on Vashon Island, Washington (Figure 2.1). Quartermaster Harbor was once rich in fish and shellfish, with abundant salmon and herring passing by the Burton Peninsula, on their return to Judd Creek and the shallows of the harbor. The point of land at the Burton Peninsula was an advantageous location for processing fish and shellfish (Figure 2.2). Not only does it allow people easy land and water access to the spawning fish, but it extends out into the windswept bay, and thus provides an environment well suited for fish and shellfish drying. Deposits of shell, charred plants, modified rocks, animal bones, and broken artifacts indicate that people caught and preserved fish and shellfish at this site. Archaeological evidence indicates that the first inhabitants occupied the point around 1000 years ago, and returned again and again in subsequent years until perhaps as recently as the 1930s. Since 1995, the waves of Quartermaster Harbor have eroded the

shoreline, including the midden, and the remains of this occupation are endangered.

The site has been the focus of many people's efforts to preserve the information it contains. In 1994, the site was officially recorded by Joan Robinson, an archaeologist working for Archaeological and Historic Research, and in 1995, it was suggested for possible nomination as a King County Landmark. The contents and age of the site, however, were unknown. In fact, no one was sure how much of the cultural material remained after the erosion caused by a 1995 winter storm.

The people working with King County's Historic Preservation Program led the preservation effort. Julie Koler (Historic Preservation Officer of Cultural Resources Division, King County Parks, Planning and Cultural Resources Department) and Holly Taylor (Heritage Program Coordinator) asked Julie Stein (then Curator of Archaeology at the Burke Museum) if the site could be excavated as part of an archaeological field school sponsored by the University of Washington. Stein

Chapter opening photo: On the first day of excavation a mother watches her child as they both learn to excavate the midden.

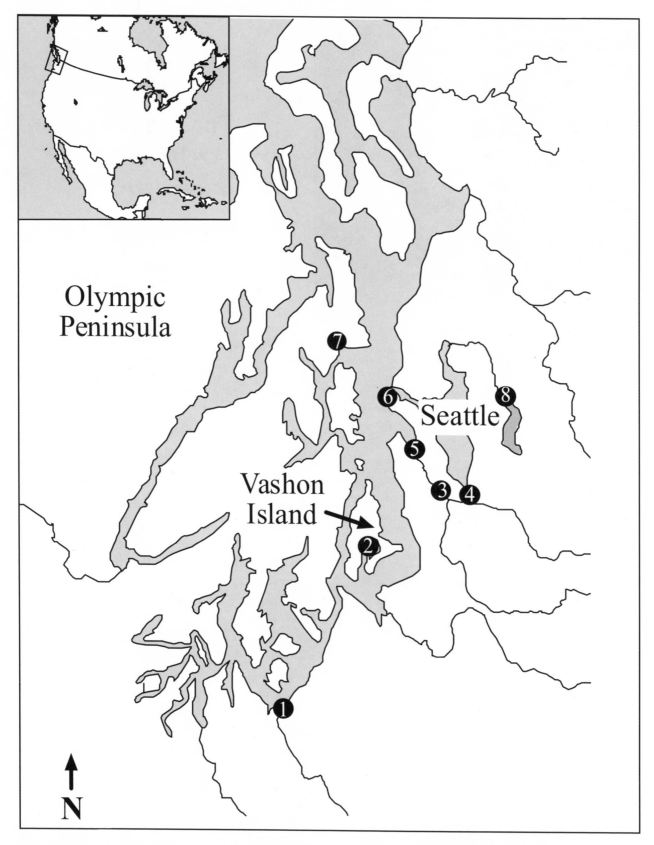

Figure 2.1 Location of the Burton Acres Shell Midden (Site 2) on Vashon Island. Sites in Southern Puget Sound mentioned in the text are 1. Dupont (45PI72); 2. Burton Acres Shell Midden (45KI437); 3. Allentown (45KI431); 4. Tualdad Altu (41KI59) and Sbabadid (41KI51); 5. Duwamish No. 1 (45KI23); 6. West Point (45KI428/429); 7. Old Man House (45KP2); and 8. Marymoor (45KI9).

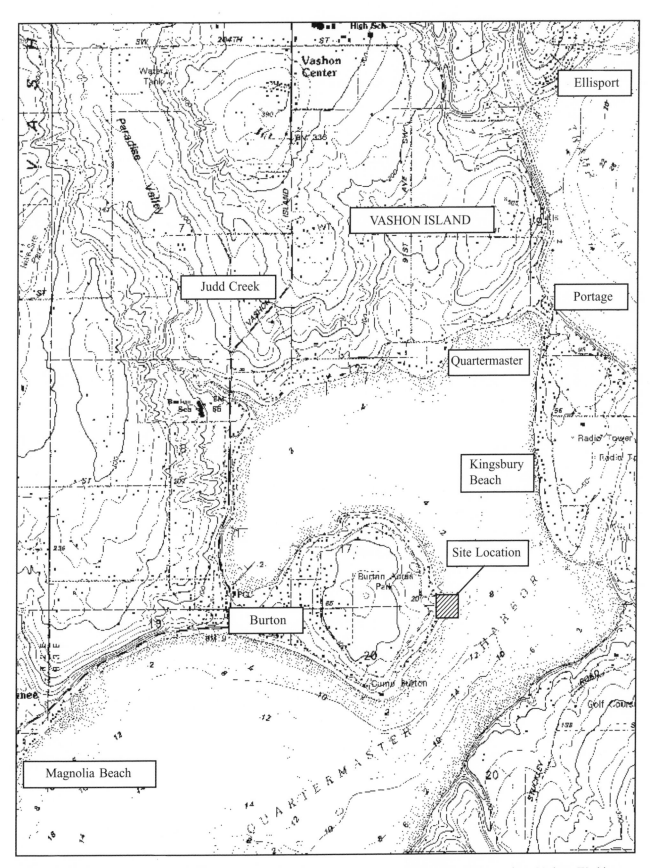

Figure 2.2 Project Area, Site 45KI437, Quartermaster Harbor, Vashon Island, Washington. (Basemap from Vashon, Washington, 1994 USGS Quadrangle).

Figure 2.3 Roxanne Thayer (right) introducing her students to the project directors, the site, and their responsibilities. Note videotaping equipment in the left background.

Figure 2.4 Visitors to the site ask questions about the excavation and obtain literature about the cultural resources from Holly Taylor (seated).

countered with the suggestion to target a wider audience through an excavation for the general public. The idea was embraced by all, but funding and permits had to be obtained.

A teacher at McMurray Middle School on Vashon Island, Roxanne Thayer, was also interested in the history of Vashon Island, and the contributions of Native Americans to the island's heritage. She was writing a grant to King County Office of Cultural Resources to make a video documentary, and wanted to include an archaeological excavation in her production. She asked Stein and Koler to include the excavation of the Burton Acres Shell Midden in her documentary project, called the "First Peoples Documentary Project". Thayer would involve middle-school and high-school students in the production of the documentary. These students would also be involved in the excavation, as tour guides (Figure 2.3). The whole project would be recorded on videotape, Vashon's & Maury Islands: Hands Across Time (Thayer 1998).

The Puyallup Tribe of Indians was immediately consulted. Quartermaster Harbor, and the Burton Acres Shell Midden, are located within the ceded territory of the Puyallup Tribe of Indians. Treaty rights granted the Puyallup people hunting and fishing privileges for this area, and their villages have been recorded in Quartermaster Harbor. Stein consulted with members of the tribe — Carol Ann Hawks (Cultural Committee), Marguerite Edwards (Repatriation Committee), and Judy Wright (Tribal Council Member) — who took the

request to the council. The Tribal Council agreed to support the project only if a Memorandum of Agreement would be signed by all parties, which stated the specific requirements insisted on by the tribe (Appendix A).

The Memorandum of Agreement stated clearly four issues agreed upon by the Puyallup Tribal Council, Puyallup Repatriation Committee, Puyallup Culture Committee, the King County Cultural Resources Division, The First Peoples Documentary Project, and the Curator of Archaeology at the Burke Museum. First, the project would be overseen by a "Project Advisory Committee", whose job would be to consider issues involving the Burton Acres Archaeological Project. The committee consisted of one representative from the Puyallup Tribal Council, the Puyallup Repatriation Committee, the Puyallup Culture Committee, the Burke Museum, the King County Cultural Resources Division, and the First Peoples Documentary Project. Second, if human remains were encountered the project would stop and the tribe would be consulted. Third, the collections would become the property of the Puyallup Tribe of Indians, loaned to the Burke Museum for the duration of the analysis. And, fourth, the publication of the results would be done in cooperation with the Puyallup to insure that both Native American and archaeological viewpoints would be represented.

Julie Stein consulted with Vashon Parks District to obtain permission to excavate, since the site is located on their property. They were also asked to consider relinquishing ownership of the collections to the

Figure 2.5 Students from the McMurray Middle School are taught how to use the screens by archaeologist Bob Kopperl (right).

Figure 2.6 Julie Stein (left) answers questions from a member of the media during excavations at Burton Acres Shell Midden.

Puyallup Tribe of Indians. Parks on Vashon Island are administered by the Vashon Parks District, not King County Parks and Cultural Resources Department (which transferred the management of most public parks on Vashon in 1995). Vashon Parks District approved the excavation and the transfer of ownership of the collection. Their justification for relinquishing ownership came after considering the expense of maintaining a collection as large as the one generated at Burton Acres Shell Midden. Annual storage and maintenance fees rise every year, and the District had no interest in saddling the community with that debt. They were willing to give up ownership in exchange for no curation costs.

Various community groups were approached to gauge local interest. Support by the Vashon-Maury Island Heritage Association was deemed essential, and was granted early in the planning stages. McMurray Middle School was suggested as a place to house the archaeological staff and the laboratory for the duration of the excavation, and Michael Kirk, the principal, agreed to this arrangement.

Funds for the excavation were obtained from many sources. Staff was funded by the King County Landmarks and Heritage Commission, King County Cultural Education Program Award, and the University of Washington (College of Arts and Science, Graduate School, Office of Undergraduate Education, and Burke Museum). Important contributions came from the Hugh and Jane Ferguson Foundation, Wyman Youth Trust, Mary and James Dunnam, Jean and Ray Auel, Vashon-

Maury Island Heritage Association, Vashon Friends of the Library, Vashon Island Rotary Club, Vashon Kiwanis Club, Deehan Wyman, Don and Jane Abel, Lizz Maunz, and Pacific Research Laboratories, Inc. This generous support paid for field equipment, rentals, and all excavation and curation supplies. Two grants from the Institute for Ethnic Studies in the United States provided money for analysis of the materials recovered from the site. Contributions from hundreds of individuals supported the radiocarbon analysis, as well as the botanical and sedimentological analyses.

SIGNIFICANCE OF THE BURTON ACRES SHELL MIDDEN

The Burton Acres Shell Midden is located in a region that archaeologists refer to as Southern Puget Sound (Campbell 1981; Wessen 1989), an area defined by access to open marine waters, sheltered marine bays, and freshwater rivers. Archaeologists, however, have excavated few locations in this area. Within Southern Puget Sound, only eight archaeological sites have been excavated that contained data sufficient to reconstruct environmental conditions and to make regional comparisons (Figure 2.1). Of these eight sites, three are littoral (located on the shore) (Old Man House 45KP2, Dupont 45PI72, and West Point Site Complex 45KI428 and 45KI429), and five are riverine sites (Duwamish 45KI23, Allentown 45KI431, Tualdad Altu 45KI59, Sbabadid 45KI51, and Marymoor 45KI9). All five of the river sites are located along the Lake Washington

Figure 2.7 The production crew from *Bill Nye the Science Guy* filmed the excavation. Bill Nye (right) watches the filming.

Figure 2.8 Bill Nye (left) answers questions from Roxanne Thayer (right) and her students.

drainage system. Due to the small sample of sites, very little is known about the archaeology of the southern Puget Sound region, and almost nothing is known about the waterways south of the Seattle area.

The Burton Acres Shell Midden is the fourth littoral site excavated in Southern Puget Sound. Its location is south of Old Man House and West Point, and its age of occupation overlaps that of Old Man House and is younger than the shell middens at the West Point and Dupont sites. Archaeologists and Native Americans know that people in Southern Puget Sound had complicated subsistence systems involving coastal and inland resources, and that the people living in Quartermaster Harbor were most certainly fishing and shellfishing there. The Burton Acres Shell Midden, however, provided an opportunity to quantify the kinds and amounts of resources taken at this one spot during this period, and to compare those resources with the discoveries at these other sites.

In addition to the archaeologists' goals, the Tribal Council of the Puyallup Tribe of Indians had a specific goal for the project. Many histories of Vashon Island have been written, and most of them start with the story of the first "American" settlers arriving on the island. The Puyallup wanted the people living on Vashon Island today to remember that the history of the island began with the Native Americans who lived in large winter villages in Quartermaster Harbor. The tribe believed that the excavation would help make that point to the community.

GOALS OF THE 1996 PROJECT
Research

As will be shown in the following analyses, many resources procured by the people who created the Burton Acres Shell Midden have been overexploited or polluted in the past 100 years, and are no longer found in the region. These data provide archaeological evidence that these resources did exist in the past and that people utilized them.

In Quartermaster Harbor interpretations of human occupation in the region differ from the ethnographic explanations because archaeologists examine the material remains (of plants, minerals, and animals) that people leave behind, while anthropologists and historians record people's ideas and language. The different observations provide different information. Archaeologists focus on people's subsistence and technology because that is what they can find. Ethnographers and historians focus on religion, ceremonies, kinship, and art, because that is what they can find. The fact that archaeologists emphasize environmental and resource factors does not imply that archaeologists believe the people occupying the location did not have complex cultures, but rather that archaeologists can best address the environmental components of those cultures. In the southern Puget Sound region the primary environmental divisions are littoral, riverine, and inland locations (Greengo 1983; Lewarch, Larson, and Forsman 1995).

Public Education

The Burton Acres Archaeological Project, publicly known as Hands Across Time, was created as a public education project. The central idea of the project was that members of the general public would actively participate in archaeology to learn more about local heritage and stewardship (Figure 2.4). Each volunteer was expected to excavate a single 2-liter bucket of material and follow it through screening and sorting, and into the first stages of analysis. Most archaeological excavations are conducted with the archaeologists in the pits doing the discovering, and the public on the outside watching. This project placed the public in the center, doing the discovering, but it also taught them in a dramatic fashion that the discovery is only a very tiny part of the process.

Volunteers who signed up to excavate were given all the equipment and forms they would need. They were accompanied by a staff archaeologist who explained and directed, but did not do the work. Volunteers made a three hour commitment; 1/2 hour was spent at the excavation unit, 1/2 hour at the screens (Figure 2.5), and a minimum of two hours at the sorting area where they separated the animal bone, shell, charcoal, metal, and chipped stone. At the sorting area, volunteers realized that they were able to do as they were directed, which allowed them to relax and ask questions about archaeology, culture history, Native American issues, and antiquity laws. After approximately three hours most volunteers had processed their 2-liter bucket and brought the sorted material to the Cataloging Station for the final check-out. Most of them returned to the beginning and took a student-led tour to see once again what they had just experienced.

Other public archaeology projects have followed a 2-week format, where a person commits to work for an entire 2-week period. Those people excavate most of the time, although laboratory work is always emphasized as well. This format has its advantages and disadvantages. It educates the people intensively and allows them to experience a sustained period of excavation. Unfortunately, this great opportunity can usually only be extended to a dozen or so individuals. Thus, the impact of such programs is powerful for the participants, but not for a very large segment of the population. A project that allowed more people to participate for shorter periods of time proved to be a very effective means of

Figure 2.9 Gene Sherman (left) shows Chester Satiacum artifacts from his properties on the shores of Quartermaster Harbor.

reaching out to the Vashon Island community and the rest of King County.

The education efforts of the Burton Acres Shell Midden excavation differed from these 2-week format projects. Work at Burton Acres Shell Midden was short and intense, but it provided ample opportunity for people to participate actively and to ask questions. People were amazed at the amount of time devoted to screening, sorting, and analyzing. They had thought archaeology was only excavation, and yet their experiences pointed clearly to the necessity of excavation being only a small part of the process. They engaged each other in lively debates on the topics of antiquity laws, traditional subsistence techniques, and Native American history. The comment heard over and over again was "I had no idea...." Many people returned later in the week to learn about any new discoveries, or to talk to people who were just experiencing the process. Often, people returned with artifacts they had found or inherited from relatives. Educational experiences were varied and numerous, and afforded ample time to learn about site protection and stewardship.

The media coverage of the project assisted the public education (Figure 2.6). Reporters from newspaper, radio, and television appeared at the site almost daily. Many broadcasts were live from the excavation. One notable event was the filming of the project for the television show *Bill Nye the Science Guy* (Figure 2.7). Bill Nye was a popular attraction. Word of his appearance spread across the island and brought hundreds of visitors to the site during the filming (Figure 2.8).

Figure 2.10 Aerial photo taken in 1936 shows the Burton peninsula before the road extended around the shoreline. Note the area of the Burton Acres Shell Midden (point of land extending to the east) with wetland nearby defined by lack of trees next to the shore.

Figure 2.11 Aerial photo taken in 1944 shows the road extending further around the Burton peninsula. Some additional trees seem to have been removed from the area around the Burton Acres Shell Midden.

The archaeological project was of short duration, spanning the period from June 21 to July 3, 1996. Each day of excavation started at 8:00 AM and ended at 6:00 PM. Visitors were scheduled in groups of three, with each group starting the process every half hour. The last group started at 2:00 PM to enable them to finish by 5:00 PM. Twelve staff archaeologists assisted the participants at the various tasks, moving from station to station throughout the day and weeks. The staff did not take days off, insuring that the site was guarded and protected.

This project reached a large number of people in a variety of ways. The general public that participated in the actual excavation numbered 375 and came from all over the Puget Sound area. Others (estimated at over 1000 people) experienced the site through guided tours as drop-in visitors from Vashon Island and the surrounding region. Residents of Vashon Island were invited to bring artifact collections to the site for identification by the Burke Museum Archaeology staff (Figure 2.9). All of these experiences were designed to educate the public about the real nature of archaeology, to preserve the eroding information at the site, to promote site stewardship, and to record information about other sites on Vashon and Maury Island (Vashon and Maury are connected by a spit of land, and are sometimes collectively called Vashon/Maury Island).

PROJECT SPONSORS AND PERSONNEL

The project, under the name Hands Across Time, was sponsored by the Thomas Burke Memorial Washington State Museum, King County Landmarks and Heritage Commission, Puyallup Tribe of Indians, Vashon Park District, McMurray Middle School, and Vashon-Maury Island Heritage Association.

The Principle Investigator was Dr. Julie K. Stein, Professor of Anthropology at the University of Washington and then Curator of Archaeology at the Thomas Burke Memorial Washington State Museum. Stein was assisted by a staff consisting of the following members: Field and Laboratory Supervisors Laura Phillips and Mary Parr, and Field Assistants Laura Andrew, Steven Denton, Michael Etnier, Steven Henderson, Robert Kopperl, Angela Linse, and Lindsay Martin. Lance Lundquist and Elizabeth Martinson, two graduate students from the University of New Mexico, volunteered for the project at the last minute. In addition to working at the site, Lundquist assisted Phillips in creating the Excel links for the database, and Lundquist and Martinson drew the stratigraphic profiles of all units. King County Landmarks and Heritage Commission and King County Cultural Education Program Award supported two staff members, Miranda Stockett and Holly Taylor. Mary Sportsman and Chester Satiacum of the Puyallup Tribe of Indians also worked

Figure 2.12 Aerial photo taken in 1960. Note that the area near the Burton Acres Shell Midden has been severely impacted with trees removed. Construction activities have extended the road all the way to the shore.

Figure 2.13 This photo, taken during the excavation and looking to the southeast, captures the shoreline along the Burton peninsula. The shore appears white because so much shell has eroded from the site.

at the site, answering questions from the public and students.

After the fieldwork was completed, several people worked at the Burke Museum's Archaeology Lab. These people assisted Laura Phillips in processing the 35 boxes of unsorted material that were excavated at the end of the last day. They were: Laura Andrew, Steven Denton, Robert Kopperl, Miranda Stockett, Sage Alderson-Gamble, Rachel Sommers, Mary Dunnam, Craig and Nancy Abramson, Fran South, Gloria Pruitt, and Pat Spier.

A group of enthusiastic middle school and high school students, brought to the project by Roxanne Thayer, guided tours and assisted visitors at the site. Many of these students also helped to create the documentary film of the project, *Vashon's & Maury Islands: Hands Across Time*. The students (in alphabetical order) were: Sage Alderson-Gamble, Annie Brule, Sonja Carlson, William Currie, Gwen Davis, Michele Guthrie, Ariel Hortz, Tegan Horan, Rachael Marcley, Timia Olsen, Erin Schlumpf, Rachel Sommers, Ben Steffen, Craig Terry, and R. J. Thomas.

The McMurray Middle School generously provided living quarters for the staff who camped in the special education classrooms. The school's cafeteria proved to be an excellent space for completing laboratory work.

THE SITE

The Burton Acres Shell Midden (45KI437) is located at the shoreline in Burton Acres Park, Vashon Island, King County, Washington. The site is a shell midden whose margins are being eroded by the waves of Quartermaster Harbor. A winter storm in 1995 caused a portion of a Madrona tree to fall and disturb a significant portion of the site. Deposits exposed in the wave-cut bank indicated that, before 1930, Native American inhabitants deposited a variety of shellfish and fish remains, charcoal, and fire-modified rock (Figures 2.10, 2.11, and 2.12). Corroded metal fragments suggested that the occupation may have occurred immediately after the introduction of metal to the local inhabitants through contact with Eurasians. The horizontal and vertical extent of the site was not known, but assumed to be confined to the immediate shoreline.

A midden is a place where people accumulate materials used to sustain life. If that accumulation includes shell, archaeologists refer to it as a shell midden. Burton Acres Shell Midden extends along the eastern tip of the Burton Peninsula, predominantly along the northern shoreline (Figure 2.13). The accumulated material includes shell, fish and mammal bone, lithics, charcoal, metal, and beach rock. This midden is unusual in two ways. First, it contains metal, indicating that the location was inhabited within the last two

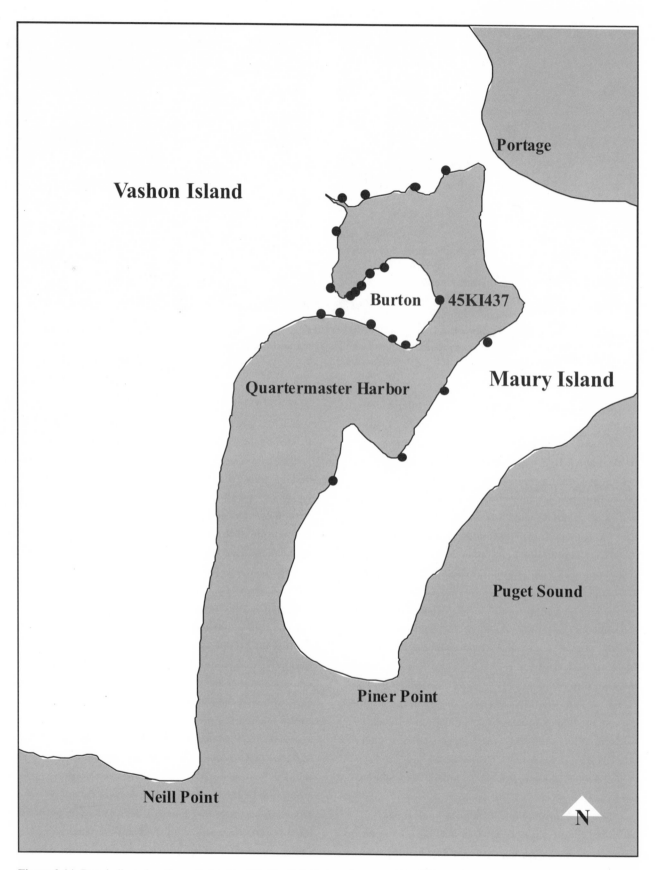

Figure 2.14 Dots indicate locations of sites identified by E.O. Roberts between 1919 and 1922. Although Roberts did not record the Burton Acres Shell Midden (45KI437); its location is noted for comparison to Roberts' sites.

Figure 2.15 Artifacts collected by E.O. Roberts between 1919 and 1922 from sites on the shores of Quartermaster Harbor. This collection is housed at the Burke Museum of Natural History and Culture.

hundred years when Europeans introduced large amounts of metal to the region. Second, the contents of the upper portion of the midden were heavily fragmented, indicating breakage and burning of shell and bone. This fragmentation could have occurred at the time of processing when the material was intentionally burned. Or it could have occurred after processing by either later occupants or even later land developers. The lowest portion of the midden differed in that it did not contain metal and did contain shell and bone that was far less fragmented. These deposits will be described in detail in Chapter 5.

Over the last 100 years, Quartermaster Harbor has been surveyed by both ethnographers and archaeologists, yet none of them noted the presence of a shell midden at Burton Acres. T.T. Waterman (1922) documents a few winter villages in the area of Quartermaster Harbor (see Chapter 3), but these records do not identify the Burton Acres Shell Midden. In 1907, anthropologist Harlan Smith noted several archaeological sites along the

shoreline of Quartermaster Harbor, but again did not include the Burton Acres Shell Midden among them. An archaeologist by the name of Earle O. Roberts surveyed Quartermaster Harbor between the years 1919 and 1922, and collected 330 artifacts from twenty sites (Figures 2.14 and 2.15), but again, Burton Acres Shell Middenwas not described in his survey notes, nor did he collect any artifacts from the site. Roberts did note, however, that the shore along the peninsula was "reported to have washed away from 10 to 20 feet in the past 15 years" (Fieldnotes on file, Accn. 2411, Burke Museum). The villages and sites recorded by these ethnographers and archaeologists still exist in Quartermaster Harbor, and landowners are aware of their presence. However, Burton Acres Shell Midden is the only site located on public land. The data collected during this excavation may not represent the kinds of artifacts preserved in those other sites, but the data do give us a glimpse into human activities in Quartermaster Harbor over the last 1000 years.

REFERENCES

Campbell, S.K. (editor)

1981   *The Duwamish No. 1 Site: A Lower Puget Sound Shell Midden*. Research Report No. 1. Office of Public Archaeology, Institute for Environmental Studies, University of Washington, Seattle.

Greengo, R.E.

1983   The Location of Archaeological Places in the Puget Lowlands. In *Prehistoric Places on the Southern Northwest Coast*, edited by R.E. Greengo, pp. 77-84. Research Report No. 4. Thomas Burke Memorial Washington State Museum, University of Washington, Seattle.

Lewarch, D.E., L.L. Larson, and L.A. Forsman

1995   Introduction. In *The Archaeology of West Point, Seattle, Washington 4,000 Years of Hunter-Fisher-Gatherer Land Use in Southern Puget Sound*, edited by L.L. Larson and D.E. Lewarch, pp. 1-33. Submitted to CH2M Hill, Bellevue, Washington. Prepared for King County Department of Metropolitan Services. Prepared by Larson Anthropological/Archaeological Services, Seattle.

Thayer, R.

1998   *Vashon's & Maury Islands: Hands Across Time*. Videotape produced by R. Thayer, Vashon, Washington.

Waterman, T.T.

1922   *Geographical Names Used by the Indians of the Pacific Coast*. American Geographical Society, New York.

Wessen, G.C.

1989   *A Report of Archaeological Testing at the Dupont Southwest Site (45-PI-72) Pierce County, Washington*. Submitted to Weyerhaeuser Real Estate Company Land Management Division, Tacoma, Washington. Prepared by Western Heritage, Inc., Olympia, Washington.

3

Ethnographic Background

LLyn De Danaan

Much of what we know about the past at the Burton Acres Shell Midden comes from the oral traditions of the Puyallup people. Historical documents for this area of Washington State survive from as early as the 1800s and record the lifeways and memories of the S'Homamish. These people are the predecessor band of the Puyallup Tribe who lived in Quartermaster Harbor and enjoyed the abundant fish and shellfish resources at Burton Acres.

"In connection with the aboriginal importance of the Puyallup, it is worth noting that they, with this Gig Harbor group (the sqwapábc including the village at Wollochet Bay and Quartermaster Harbor) commanded the only water entrance to the entire southern section of Puget Sound."
— Marian Smith[1]

"They had many buildings at Quartermaster Harbor, at portage, and down on the entrance of Quartermaster Harbor. ...At the entrance of Quartermaster Harbor on Maury Island was a building there; it was built by my people...200 feet long."
— Lucy Gurand, Puyallup/*S'Homamish* 1927[2]

This brief essay will serve as an introduction to the people who demonstrably occupied Vashon Island during the early middle and middle of the nineteenth century. We have no reason to doubt that these people

were descendents of those who left their mark on the island from a much earlier period.

The inhabitants of the island, called the *S'Homamish*, were a Puyallup Tribe predecessor band and occupied several villages and houses on and around Vashon Island, including Gig Harbor and Wollochet.[3] *S'Homamish* (also referred to as the *Sqopabc* and *Swapabc* in Smith) lived on Vashon Island and across West Passage to the west of Vashon in the areas of Gig Harbor or Wollochet.[4] This band appears in the literature under various spellings or names, some apparently referring to specific villages. Some of these names or spellings include: *Skwapa'bsh*,[5] *S'Ho-ma-mish*,[6] *Homamish* (in *Indian Statistics Within the Limits of the Second Sub Indian Agency of Oregon Ter.*, a report written by Anson Dart, Superintendent of Indian Affairs, 1851[7]), and *Shone mah mish* (in the *Census of Indian Population in the Fort Nisqually District as taken in the years 1838-39*[8]). They were members of the

---

Salish speech community, a community that included predecessors of the current day Squaxin, Nisqually and Puyallup people. They built localized cedar plank houses, which were occupied in the winter and were basically independent social units.

Though only fragmentary information exists that describes and defines the *S'Homamish* specifically, it is clear that these people shared in the main cultural patterns of other predecessor Puyallup groups and Southern Coast Salish people in general. There is considerable evidence of their occupation of the island, including census reports, reports of informants who worked with anthropologists and archeological survey-ors, reports of Hudson Bay Company and tales collected by people such as Arthur Ballard. The testimony of Lucy Gurand, who was about eighty-five at the time of the Court of Claims,[9] stands as witness to the viability of the villages and village culture that she knew as a young woman. She would have been around twelve years old at the time of the signing of the Medicine Creek Treaty in December of 1854.[10] There are reports of Indian houses around the Sound and Hood Canal being destroyed or burned around the mid-1860s, that is, ten years or so after the treaty and after people were coming in to the reservation at Puyallup. Thus, Lucy Gurand may have seen these *S'Homamish* houses still standing into her teenage years. Lucy Gurand was also an informant to T.T. Waterman,[11] living on Vashon when Waterman was collecting place names. In fact, she apparently lived there at Burton,[12] possibly to the time of her death and was well known to other Vashon Island residents as a friend, a fine knitter, and a clam digger. Waterman notes that her grandfather had been "headman at Quartermaster harbor."

Waterman reported that his informants did not have names for whole land masses such as islands.[13] He believed that people of Puget Sound related to the coastline. For example, there was no name for what is now called Bainbridge Island, but there were "upward of three hundred names for different places on the island. If an Indian in a canoe were 'headed,' as he would say, for Bainbridge Island and were asked where he was going, he would in reply name the *spot* where he proposed to land."[14] Native people named specific features of historic, navigational and ecological/subsistence significance to them.[15]

The record contains many names that mark features along the coasts of Vashon and Maury Islands.[16] These names indicate intimate familiarity with the geography of the area and extensive use of the islands over time. Some names act as mnemonics for legendary acts or persons such as [Tu]*ksiu[e]b*, "where snakes landed." Others recall resource-rich areas such as Peter Point or "Qo':ati, where cattails are found." Still others are cautionary or provide navigation points for sailors approaching the island.

A number of the Vashon Island place names recorded by Waterman indicate places of habitation or long-term use as spring or summer camps. These include Waterman's #201 *Qw[3]Eq[3]ks*, "white promon-tory," a place where Tom Gerand (Gurand in other texts) lived, which was "covered with broken shells; #203 "an ancient site with a kitchen-midden one foot deep, QoqóLttcEtc," one of the places Lucy Gurand recalled seeing houses when she was young; and #204 "an old village site with three feet of kitchen refuse, kwll[3]ut." This is the site, according to Waterman's notes, of the "war with the snakes." Another village site listed by Waterman is #208 *A'lalEl*.

It is perhaps Lucy Gurand, whose image appears in a faded copy of a Tacoma newspaper from the time of her court appearance (Figure 3.1), who has given us the richest insight into pre-European and Euro-American life on Quartermaster Harbor and at Portage, as she responds through an interpreter to questions put to her in a 1927 Court of Claims deposition.[17] In the newspa-per account, Lucy Gurand (called Lucy Slagham in the article) is pictured with Wapato John, a Puyallup allottee, and Tom Milroy (one of Arthur Ballard's informants for *Mythology of Southern Puget Sound*), two other Puyallup witnesses to the Court.

Wapato John, whose age is given as "somewhat over eighty,"[18] in his testimony in the Court of Claims two days later, affirms that Lucy Gurand's description of the buildings at Vashon was "about right."[19] There is another exchange about Vashon later in Wapato John's cross-examination:

"Question: Now I want you to answer this question "yes" or "no." Was Maury Island in the Puyallup country?

Answer: Yes

Question: Was Vashon Island in the Puyallup country?

Answer: Yes"[20]

Figure 3.1 Lucy Gurand is shown in this March 25, 1927, newspaper photograph. This photograph was published during the time of her Court of Claims deposition. She is standing with Wapato John (left) and Tom Milroy (right), two other Puyallup deponents. All three were living when the Medicine Creek Treaty was signed. (Used with permission of the Tacoma Public Library and Tacoma News Tribune).

By the time of Wapato John's testimony in 1927, the houses were gone. Even ten or so years earlier when Waterman visited the island with his informants, he saw only remnants of middens, and noted in some cases that "thirty feet of land have washed away" where house-pits had been. It was Lucy Gurand and Wapato John and a very few others who could attest to life as it had been before the influx of American settlers dramatically and permanently changed the *S'Homamish* people's way of life. However, it is clear from the record that not all of the *S'Homamish* left their traditionally held territories and moved to the Puyallup reservation.[21]

John Xot,[22] another Puyallup allottee, in conversation with Arthur Ballard,[23] recalled a village south of Burton on Quartermaster Harbor (cf. Waterman) and one at the mouth of the bay at Gig Harbor.[24] The people of the Vashon area were called the "swift water people," according to Joe Young,[25] a Puyallup allottee who worked with Ballard and Waterman. This appellation was apparently in recognition of their skill at navigation on the deep waters of the Sound as well as a description

of their home across from the Narrows.[26] In general, the *S'Homamish* were clearly "salt-water" people as defined by Smith. Joe Young also told Ballard a number of place names on and around Vashon and mentioned a house at Clam Cove with a "depression," i.e. that its foundation was dug into the earth.[27]

Because elders such as these provided depositions to the Court of Claims and worked with anthropologists such as Waterman and Ballard, we have a foundation for placing the *S'Homamish* in space and time based upon memory that stretched back prior to treaty time (1854). This information, along with that available from primary historical sources such as notebooks, letters, journals and official reports of Europeans and Euro-Americans such as Gibbs and Tolmie,[28] later oral and written contributions by Puyallup people such as Silas Cross and Hattie Cross, and studies by anthropologists such as Marian Smith (who worked with Puyallups William Wilton, Mary Anne Dean, John Milcane, Annie Squakwium, Annie Squally, Jerry Meeker, and Joe L. Young), and Hermann Haeberlin and Erna Gunther (who worked with Puyallup allottee Henry Sicade)[29] allow us to paint a rich and multifaceted, though necessarily incomplete and certainly fragmentary, picture of the political, cultural, social and economic context in which the *S'Homamish* people lived.

In the following pages, we will provide a general introduction to the ethnography of the Puyallup people and their predecessors, including the *S'Homamish*. This introduction will include a discussion of geographic distribution and general ethnographic information, including house and village types and locations, subsistence strategies, technology, and social structure. We will not pretend to be holistic in our presentation here. But we will attempt to provide a broad ethnographic and historic context as a basis for further study of the people who were among the predecessors to the present day Puyallup.

After establishing a general ethnographic framework for the Puyallup, we will discuss the *S'Homamish* of Vashon Island, Gig Harbor and Wollochet. We will review some of the evidence for their occupation of Vashon, including population estimates from periodic early to mid-nineteenth century census reports, notations on mid-nineteenth century maps, and written comments regarding Vashon from reports, letters or essays written in or around this mid-nineteenth century period.

GENERAL ETHNOGRAPHIC CONTEXT: THE PUYALLUP

The present day members of the Puyallup Tribe are descendents of Southern Coast Salish people who lived along the drainage of the Puyallup River, and adjacent to waterways in Southern Puget Sound.

The Puget Sound was not only valued by the indigenous Puyallup predecessors but also quickly became a main lure to Euro-Americans, who, in the mid-nineteenth century, saw the "towering forests rising above deep water anchorages" as perfectly suited for commerce.[30] "There is no country in the world that possesses waters equal to these," Thomas Farnham wrote during this expansive period in United States History.[31] And Charles Wilkes of the 1838-42 United States Exploring Expedition predicted that the "future state is admirably situated to become a powerful maritime nation."[32]

Within a few years of the United States' treaty with Great Britain in 1846, the conclusion of the so-called Stevens' treaties (1854-1856) with the indigenous people of Washington, and the influx of Euro-American settlers, life on Puget Sound was inevitably altered.

The orientation to Puget Sound and the rivers and waterways that drain into it was, before this demographic, cultural and economic transformation, the most obvious determinant of the shape and content of Puget Sound cultures. Marian Smith wrote:

"...villages were placed either on small streams which tumbled down to the Sound from the high lands of the Kitsap Peninsula, or along the river system which drained the west slope of Mt. Rainier and the country lying between the mountain and Commencement Bay. The waters of these streams flowed in a radial network, all of them pointing eventually toward a hub located at or around the mouth of the Puyallup River, the present site of Tacoma, just south of which the Sound narrowed forming a bottleneck entrance to its upper reaches...."

"It was a land with a heavy rainfall, a temperate land of swamps, extensive tide flats, damp, overgrown gullies and sudden floods. The bed of a stream might shift fifty feet in a single year only to return to its old channel the following spring. Salmon were plentiful and could be caught in the Sound at any time of year..."

"Open country was apt to be only along water courses swept by tide and flood, and each village was set at the mouth of a stream where it entered the Sound or, in the case of tributaries, at its junction with another stream."[33]

The Puyallup River runs from the southwest slope of Mt. Rainier, specifically from the Klapatche Ridge area, and enters Puget Sound at Commencement Bay. The people of the Puyallup drainage system, or the *Spwiya'laphabac*,[34] could travel down stream or overland to Commencement Bay by trail or shovel-nosed canoe. The people from Commencement Bay could travel easily to Vashon Island and Quartermaster Harbor in large canoes designed for the waters of the Sound. The people from Vashon, accustomed to the rough, open water of the Sound, and possessors of big saltwater canoes for hauling people, freight or for fishing,[35] could travel easily around the Sound through the Narrows to Wollochet or into the calmer waters of Commencement Bay. Archibald Menzies, the botanist who visited the Sound with George Vancouver in 1792, and Charles Wilkes, who sailed around Commencement Bay and visited Fort Nisqually during his 1838-42 United States Exploring Expedition, both noted the navigational challenges of the Narrows. By implication, the Puyallup predecessors who traveled these waters were skillful sailors.

If orientation to the water is the most obvious determinant of the culture, then geographical difference in how and where one primarily interacted with water was the primary determinant of differences between and among people. In notes for *Native Geography in the Puget Sound Region*, Waterman wrote, "I would say that the difference between salt-water people and fresh-water people is the one ethnographic distinction which rises most prominently in the Indian's mind even today. ...There are certain constant differences in culture between the people living "on salt-water" and those living in the interior which justify the distinction."[36]

Smith noted the general differences among and between Puyallup predecessor bands. Their relationship to the river or saltwater, and their proximity to other resources differed considerably.

"*sxwaldjabc*. People of the salt water. These groups lived on the Sound and they were characterized as canoe Indians. They possessed canoes capable of navigation in the rough waters of the Sound and were skillful in the handling of such craft."

"*stologwábc*. River people. A name applied by peoples located on the Sound to groups above them on the same river drainage. It implied that such people were comparatively unfamiliar with the Sound and navigation upon it. They were a particular kind of salt water people."

"*tálebiuqᵘ*. Inland people. This term was used in contrast to that for salt water people. These inland groups traveled back and forth above the river beds in the country paralleling to the Cascade Range."

"*sabákwebabc*. Prairie people. These were characterized as horse Indians. They were inland groups of a particular kind, the differentiation resting upon their ownership and use of horses."

"*sták'tabc*. People who had no waterfront connections, or, more accurately, who lived on rivers which drained away from the Sound."[37]

The Puyallup predecessors were Salish speaking people. Bands and villages of Puyallup predecessors were linked not only by language but also by marriage, periodic occasions for feasting and other activities, and sharing of some fishing, clamming and other resource gathering territories. Smith writes, for example, that though meetings between hunters might be sporadic or infrequent, people often anticipated annual encounters at particularly good clamming grounds or productive root or berry grounds.[38]

Decisions of where to live seemed to be at least in many instances based upon an assessment of availability of resources as well as the presence of someone already in the household with whom one could claim kinship. Decisions to stay or move were often very practically considered ones.

Multifamily houses were common and sometimes several of these were found in close proximity around the mouth of a river or stream. Villages were mainly composed of a number of houses with persons related to one another by blood or marriage. But occupation of a house was not restricted to related persons. According to Smith, unrelated individuals were never excluded for that reason alone.[39]

Kin groups spread over a large geographic area, and efforts were expended to increase that area by marriage with distant households. Villages were exogamous. Kinship was reckoned bilaterally, and matrilocal and patrilocal residence was fairly evenly distributed.

Once a family or individual was established as a member of a house, one had a designated space within the house. Each family group had an area of its own with a fire, sleeping benches, and storage room. Thus, a household consisted of several family groups. House groups engaged in some cooperative enterprises, such as the building of fish weirs and nets, and had persons who were specialized in some aspect of production. Similarly, house specialization applied to various activities, such as canoe building, basket making, or hunting. The house group was always in some stage of a "developmental" cycle. That is, new members might be joining the group, or persons might be marrying and moving to another household.

Village groups consisted of households in geographic proximity and were composed primarily of persons born and raised within the village regardless of blood ties. Those most familiar with the rivers, the land, and the resources of the area tended to be most identified with the area. They carried the stories, the history, and the local culture. These houses and villages were built along the shore and faced the water. The houses might be as long as one hundred to two hundred feet.[40]

Smith, on the basis of her interviews, concluded that an average sized family group was about five and that villages were composed of twenty-five to fifty households. Lucy Gurand's deposition in the 1927 Court of Claims, *Duwamish et al v. U.S.A.*, offers a perspective on those houses and their sizes on Vashon Island as she remembered them from the treaty period:

"Question: Tell us the villages and the number of houses, as much as you can, of the Puyallup people, not including those on the Puyallup Reservation.

Answer: They had many buildings at Quartermaster Harbor, at portage, and down on the entrance of Quartermaster Harbor.

Question: How many at Quartermaster Harbor?

Answer: At the entrance of Quartermaster Harbor on Maury Island was a building there; it was built by my people as a fort...200 feet long.

Question: How wide was it?

Answer: In the neighborhood of 50 to 60 feet wide.

Question: Did the people live in that?

Answer: Yes, sir; every winter.

Question: Were there any other houses there at that place?

Answer: There were houses further above the harbor.

Question: How many there?

Answer: Four houses, small houses, above that fort.

Question: How big were they?

Answer: Ranging from 40 to 60 feet.

Question: And how wide?

Answer: Thirty to thirty-five feet, probably.

Question: What other houses were there?

Answer: At the portage there were quite a number up there.

Question: How many there?

Answer: She said there was quite a number of them. About seven families owned this building, at Portage; that is, in each, each building.

Question: And she held up her six fingers. Were there six buildings there?

Answer: There were seven buildings and about seven families in each.

Question: How big were those seven buildings?

Answer: They ranged from 45 to 50 feet long.

Question: And how wide?

Answer: Thirty to thirty-five.

Question: What other place were there buildings?

Answer: The same band of Indians had buildings at Gig Harbor."[41]

The Puyallup proper were, according to Smith, located at the mouth of the Puyallup River, with house sites at the present 15th Street and Pacific Avenue. Waterman called this a "large and important village." However, the term was extended to include three villages immediately contiguous to it named, *twadebcab*,[42] *catcqad*[43] and *kalkalaq*.

Aside from the rivers, the main geographic features of the area are the Puget Sound basin itself with the long trough of Puget Sound terminating at its southern extremes in a series of long narrow inlets, the Olympic Mountain range to the west rising to 8000 feet, and the Cascade Range on the east, with the prominent volcanic peak Tahoma (Mt. Rainier), a significant part of the cultural as well as geographic landscape. This rich ecosystem included ancient forests of western hemlock (*Tsuga heterophylla*), Douglas fir (*Pseudotsuga menziesii*), and western cedar (*Thuja plicata*), as well as large leaf maple (*Acer macrophyllum*), alder (*Alnus rubra*), madrone (*Arbutus menziesii*), ash (*Fraxinus latifolia*), yew (*Taxus brevifolia*), and garry oak (*Quercus garryana*) which produces edible acorns.

Within the region were large prairies with widely spaced trees. These prairies and some open woodlands were clearly maintained by periodic burning that enhanced production of plant food such as huckleberry and root crops.[44]

The extensive list of floral resources used in this region, known as one of the richest and most productive ecosystems on earth, includes salmonberry (*Rubus spectabilis*), thimbleberry (*Rubus parviflorus*), blackcap (*Rubus leucodermis*), wild strawberry (*Fragaria cuneifolia*), gooseberry (*Ribes divaricatum*), red-flowering currant (*Ribes sanguineum*), salal (*Gaultheria shallon*), red huckleberry (*Vaccinium parvifolium*), and cranberry (*Vaccinium oxycoccus*). People also used licorice fern (*Polypodium*), sword fern (*Polystichum munitum*) and other members of the Polypodiaceae family especially the rhizome of the bracken fern (*Pteridium aquilinum*). Other common food plants were members of the Liliaceae family including camas (*Camassia quamash*) and tiger lily (*Lilium columbianum*).[45] Some of these root vegetables were staples of the pre-European American diet. Plants provided not only food but also material for housing and for the manufacture of hunting and fishing technologies, and medicinals.

Most important to the lives of Puget Sound people were finfish, especially the anadromous fish including the tyee or Chinook salmon (*Oncorhynchus tschawytsca*), the silver (*O. kisutch*), dog (*O. keta*) and humpback (*O. gorbuscha*), and steelhead ocean-going trout (*Salmo gairdneri*). Other seafood commonly consumed included bullhead, rock cod, devil fish, eel, flounder, perch, smelt and herring. Devil fish were considered a favorite, according to Smith. She reports that people of the upper Puyallup valley traveled to the Sound to the area of Redondo Beach to catch them while they were "asleep" on the shore.[46]

Saltwater fishing techniques employed by Puyallup predecessors included fishing from a canoe with long line and hook, trolling from canoe with hook, dip netting herring from the shore, and spearing and raking herring in deep water. The Puyallup also practiced seine fishing with netting "as much as two hundred feet in length and from six to seven feet wide." River fishing technology included elaborate traps and weirs placed in the rivers. Puget Sound people also used dip nets and spears on the rivers.

Shellfish were also abundant and were collected seasonally. Shellfish commonly taken for food included oysters, littleneck clam, butter clam, bay mussel, horse clam, cockle, and geoduc. Sea cucumbers and sea eggs were also named as important and favored foods by Marian Smith's informants. Shellfish, especially clams, were commonly dried for winter use.

Ballard reported that people living inland took account of the moon's phases and their relationship to the tides. Thus, people timed their visits to the shore in order to take advantage of extreme low tides favorable to the gathering of shellfish.[47]

Shellfish were among the resources gathered during summer expeditions. Clam digging, according to Joe Young in conversations with Ballard, was best begun in the spring when "the dogwood is in bloom." Meetings of friends and kin occurred at exceptionally good clamming grounds and places that produced roots and berries. These meetings were informal but "annual, anticipated occurrences."[48]

Vashon Island is included in Smith's list of seasonal harvest regions. Jerry Meeker reported that "Quartermaster Harbor was the annual place for clams. Puyallup, Gig Harbor held camp at Pt. Defiance and went across to Q.M. for clamming."[49] Wapato John named other camp areas on Vashon Island, including Clam Cove.

Land mammals were also a source of food. Deer and elk were hunted with bow. These, as well as bear, beaver and other smaller mammals, also were taken in pitfalls and snares. Snares were set up along the runways of the animals. Hunters set multiple nooses along the path so that it was almost impossible for the animal to escape them all. Some ducks were also caught with nooses, though these and other birds were also speared or caught in large aerial nets.

THE S'HOMAMISH

Early European and Euro-American journal accounts of Indian life in and around the territory occupied by the *S'Homamish*, though intriguing, are anecdotal and superficial and, therefore, do not add much to the record. For example, Peter Puget encountered native peoples at Olalla in May of 1792. Olalla is across West Passage from Vashon, about midway up the West Passage. John Work made note of a camp (which journal editor Elliot believes might be Vashon Island) in his journal entry of December 7, 1824: "Where we camped is an island, where we see the marks of some horses which the Indians have on it."[50]

Horses were owned by Puyallup-Nisqually Indians during this period, according to the 1838 Hudson Bay census. In his October 18, 1835, journal entry, Work camped opposite the north end of Vashon Island. Though he reports passing a "camp of Indians" in the evening, it is not clear exactly where they were.[51] Since he was traveling south towards Nisqually, the people he saw could have been resident in the land on either side of the West Passage. Augustus Kautz wrote that the Narrows were alive with "Indians fishing for small fish and salmon. They trolled for the salmon and caught the small fish with a species of rake...." as he approached his camp on Vashon's Island on May 23, 1853.[52]

The "official" historical record, however, is quite clear regarding the identity, home territory, and vitality of the people who occupied Vashon Island at, and before, treaty time. This record includes censuses, maps, official correspondence and the Medicine Creek Treaty itself. The *S'Homamish* band is named in the preamble of the Medicine Creek Treaty:

> "Articles of agreement and convention, made and concluded on the She-nah-nam or Medicine Creek, in the Territory of Washington, this twenty-sixth day of December, in the year One thousand Eight hundred and fifty four, by Isaac I. Stevens, Governor & Superintendent of Indian Affairs of the said Territory on the part of the United States and the undersigned Chiefs, Head Men and Delegates of the Nisqually, Puyallup, Steilacoom, Squawksin, S'Homamish, Steh-chass,

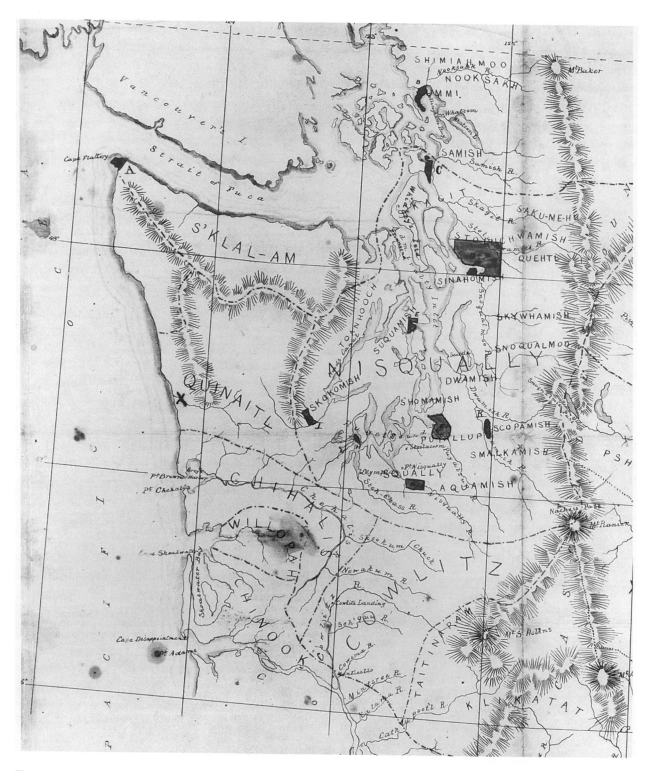

Figure 3.2 This 1854 "Map of Washington Territory Showing the Indian Nations and Tribes" was produced from data compiled by the Pacific Rail Road Expedition and Survey. The map indicates that the S'Homamish occupied Vashon and Maury Islands. (Used with permission of the Washington State Historical Society, Tacoma, Washington, Negative Number 1990.60.1).

T'Peek-sin, Squi-aitl and Sah-heh-mamish Tribes and Bands of Indians, occupying the lands lying around the head of Pugets Sound and the adjacent Inlets, who for the purpose of this Treaty are to be regarded as one Nation, on behalf of said Tribes and Bands and duly authorised by them."

The Treaty was signed on December 26, 1854. In a letter dated January 6, 1855, George Gibbs wrote to Territorial Governor Stevens that the *S'Homamish* wanted their own reservation in "their country." Gibbs reported to Stevens that it was "too late" and that this request should have been "mentioned before signing the paper."

Earlier, an 1838-1839 Hudson Bay census reports the total population of the *S'Homamish* band at 315.[53] The *S'Homamish* are recognized in an 1849 letter to the Office of the Superintendent of Indian Affairs written by Joseph Lane.[54] He reports that, "the Homamish, Hottunamish, Iquahsinawmish, Syhaynamish and Stitchafsamish indians ocupy [sic] the country from the narrows along the western shore of Puget's sound; friendly and well disposed; Total about 500; subsist by labor and fishing." In an 1851 census taken by Anson Dart (*Indian Statistics Within the Limits of the Second Sub Indian Agency of Oregon Ter.*) the *S'Homamish* are grouped together with other bands "from the narrows, along the western shore of the Pugets Sound to New Marke [sic]."[55] In a September 1, 1852 letter from E.A. Starling to Anson Dart, Starling comments that he "intended to prepare a map of the area," and had not. However, he does forward the names and locations of the tribes. These include the "Sho-mam-ish, Vashon's Island."

In February of 1854, W.F. Tolmie of the Hudson Bay Company provided George Gibbs with a census of "Various Tribes Living on or near Puget Sound" taken in autumn of 1844.[56] The *S'Homamish* are reported to have a population of thirty-four men, twenty-two women, thirty-four boys and twenty-eight girls. They are reported by him to have thirty-four canoes. In this same letter, Tolmie reports a current (1854) census. This one reports the *S'Homamish* with a population of eight men, seven women and eighteen children. George Gibbs in his March 1854 study called *Indian Tribes of Washington Territory* addressed to Captain McClellan reports on the population of the *S'Homamish* "of Vashon's Island." He reports that a total of eighteen men and fifteen women have been counted.[57] The *S'ho-ma-mish* are also in Gibbs' notes of a February 1854 census.[58]

A map of Washington Territory, dated April 14, 1854, was drawn by a topographer for the Pacific Railroad expedition (Figure 3.2). The "Indian names and boundaries" were placed on the map by George Gibbs. The location of Indian people represented by this map is the result of Gibbs' research and data collection. The "Shomamish" name is inscribed over Vashon Island.[59] A second map dated 1854 (but clearly made after the treaties in Washington were signed) also shows the *S'Homamish* across Vashon Island as well as Maury Island and part of the Kitsap Peninsula.[60] The *S'Homamish* are placed on Vashon in subsequent maps such as the "rough tracing" map transmitted (with the Medicine Creek Treaty) by Isaac Stevens December 30, 1854, to George Manypenny, Commissioner of Indian Affairs, in Washington, D.C.[61] "Shomamish" is written over Vashon Island in a variant map of the Treaty area, also dated December 30, 1854.[62] Prior to the signing of the treaty, Michael T. Simmons apparently provided Gibbs and Stevens with census estimates which Gibbs notes, "are considerably altered by the subsequent census." In the report, Simmons lists "Puyallups & S'Homamish" together with head chief Kwa-ta-hute. This report is apparently in response to a March 22, 1854 letter from Isaac Stevens to Col. M.T. Simmons in his role as Special Indian Agent.[63] In this letter, Simmons was instructed to establish an agency and travel among people, gathering bands into tribes and appointing chiefs and sub chiefs to prepare for treaties. "Pick people who will control them to their best advantage," the instructions read.[64]

In 1855, George Gibbs produced yet another map depicting territory ceded in the Western Washington treaties. Again, the term *S'Homamish* is clearly written over Vashon Island and adjoining land mass to the west of the island.[65] We have attempted to reproduce the editorial marks and revisions in a manuscript draft version of an 1856 report to Isaac Stevens in which Gibbs wrote:

"There remains on these waters what may be termed the <u>Nisqually nation</u> which is thus divided. 1st The bands occupying Pugets Sound proper & the inlets opening into it extending to and including the Puyallups & S'Homamish. ~~The bands constituting~~ These all speak the same dialect were included in the treaty made at Shenahnum or Medicine Creek in December 1854, which has since been ratified. They number collectively about ~~640~~ 893. A division might be made of these into three tribes, the first including all the bands ~~west of~~ from the mouth of Nisqually river westward ~~indecipherable~~ who are properly salt water Indians…"

In the same draft document of the report to Stevens, Gibbs specifically mentions the *S'Homamish* as part of what he labels a third division, "~~including~~ consisting of the ~~Puyallup and the S"H~~ Indians ~~inhabiting~~ living on the Puyallup river and ~~Vashons Island, who~~ the S'Homamish of Vashons Island, <u>who are very intimately connected</u> with them. [66]

In Gibbs' 1877 *Tribes of Western Washington and Northwestern Oregon*, he wrote:

"The bands occupying Puget Sound the inlets opening into it as far down as Point Pully. These all speak the same dialect, the Niskwalli proper, and were all included in treaties made at Shenah-nam, or Medicine Creek, December, 1854, since ratified by the Senate. They number collectively 893. A division might be made of these into three subtribes, the first consisting of the S'Hotlemamish of Case Inlet, Sahehwamish of Hamersly Inlet, Sawamish of Totten Inlet, Skwai-aitl of Eld Inlet, Stehsasamish of Budd Inlet, and Nusehsatle of Sourth [sic] Bay or Henderson Inlet; the second consisting of the Skwalliahmish or Niskwalli, including the Segwallitsu, Steilakumahmish, and other small bands; the third of the Puyallupahmish, T'Kawkwamish, and S'Homamish of the Puyallup River and Vashon Island."[67]

The last division, he asserts, are River and Sound Indians.

CONCLUSIONS

The people known as the *S'Homamish* occupied Vashon Island and the surrounding area during and before treaty time. They were saltwater people, skillful sailors and intimately associated with other Puyallup predecessor people, especially people who occupied households and villages around Commencement Bay. They shared generally in the cultural practices established among the Southern Coast Salish people, including joining others in seasonal food gathering camps or temporary villages, and hosting friends and kin on special occasions. They enjoyed the abundant finfish and shellfish resources from the Sound as well as the vast root and berry crops. Their diet included land mammals of many varieties, and also birds and ducks. They lived in large multiple family cedar plank houses which contained built-in sleeping platforms, storage space, drying racks, and fire hearths. They shared stories that featured local landmarks and regional histories. Their households functioned as did those of other Puyallup predecessor people; as economic units that shared the production of members, especially by those skilled or talented in basketry, hunting, and other special professions. Leadership in interpersonal relations was valued and reputed beyond the village level.

The *S'Homamish* were among those party to the Medicine Creek Treaty of 1854. Some *S'Homamish* people, especially those at Wollochet Bay, though allottees and enrolled at the Puyallup Reservation, remained in their home area for many years after the Treaty was signed.

ENDNOTES

The Puyallup tribe and predecessor bands have been the subject of a number of anthropological studies over the years, many of which have been cited above. The earliest formal studies might be said to date from observations collected by George Gibbs and Dr. W.F. Tolmie. Franz Boas, often called the "father" of American anthropology, visited the Puyallup Reservation in the summer of 1890. From Victoria, B.C. he wrote to his "dear wife" that he had "measured 35 people"[68] while visiting the Puyallup. According to Edwin Chalcraft's personal memoir, Boas was engaged in measuring "full-blood" Indians for the Smithsonian Institution.[69] Among others who used Puyallup informants during succeeding years was John Winterhouse, Jr., who consulted with Jerry Meeker for his archaeological survey of lower Puget Sound in the 1940s.[70] Melville Jacobs, along with noted ethnomusicologist George Herzog and Marian Smith, recorded songs and some stories with Puyallup Jerry Meeker in 1951.[71] Arthur Ballard, already mentioned, worked in Auburn, primarily with people on the Muckleshoot Reservation, but a number of enrolled Puyallups and people from Puyallup predecessor bands were among his informants.[72] Hermann Haeberlin and Erna Gunther's slim volume on Puget Sound Indians[73] notes that Henry Sicade was an informant. Haeberlin worked with "the oldest people available" in 1916-1917. The most complete and widely referenced Puyallup ethnography is that written by Marian Smith based upon her work with informants from October 1935 to May 1936.

1. Marian W. Smith. *The Puyallup-Nisqually*. Columbia University Contributions to Anthropology 32. New York. 1940. (Reprinted: AMS Press, New York, 1969). P.11.

2. This quote is from Lucy Gurand's deposition in 1927 during Duwamish et al. v. U.S.A., also referred to as Court of Claims. (Court of Claims of The United States. No. F-275. Duwamish et al. v. U.S.A. 1927).

The context of this quote appears later in this essay. Gurand notes that the building is a "fort." Fort-like structures were not unusual in Puget Sound: "Sometimes the village was protected by a palisade of cedar, about fourteen feet high. Rocks were piled up at the base of the logs that were set about three feet into the ground. On top of the upright posts was laid a horizontal log grooved to fit the top of the uprights and tied to them with cedar rope. The palisade had doors of solid cedar planks barred on the inside with two horizontal beams. In addition to the doors, there were openings in the wall about one yard square for shooting arrows. These were cut at the height of a man's shoulder. These could also be closed from within by cedar boards that were bolted. These palisades enclosed the village on all sides including the water side" (Hermann K. Haeberlin and Erna Gunther, *The Indians of Puget Sound*. University of Washington Publications in Anthropology 4(1):1-83. Seattle. 1930. P. 15). Smith also notes that *tsugwᵻlEl* was originally a "fortification." Smith notes that this village was built by a single man who was in danger from the Duwamish. Arthur Ballard collected a story that may relate to the necessity of having a fortified building. The story, in this version, was told by Big John from Green River:

"A young man of the sxwababc lived on Kwilu't on what is now known as Quartermaster Harbor in the southern part of Vashon Island. Once he sought a wife. To the village of Staq on White River he came and there he passed many days, but no wife did he find. At last, giving up the search, he returned to his home across the

Sound. Now while on his visit to White River the young man had killed a very handsome garter snake. It so happened that this snake was the son of the chief of the snake people and the snake chief was angry. So the snake chief gathered his people together in council and said, "Let us go to the village of the sxwababc and there destroy them; let us make war upon them." It was agreed. All of the snake people began the journey. At White Rock near the prairies they came to the bay, near the present site of Des Moines. Out in the bay they spied a fisherman in a boat. They hailed him and bade him carry them across in his boat. But the boat was too small for so many people. So the fisherman let trail in the water a long rope which was attached to the stern of the canoe, and all the snake-people laid hold on the rope until it was full for the entire length, and the boatman towed them across to the place where a cliff overhangs the water. Early in the morning they approached the young man's village. A lone woman dipping water in a basket espied the attacking party and ran to all the houses crying, "The snake people are coming! They are numerous!" Then the snake people attacked all the people wherever they found them and in whatever manner they could reach them. "Hadeda! ha-ada-a-a-da!" There is another one, the snakes would cry, as they saw the people in their houses. Thus they continued till all their enemies were destroyed and they were avenged for the death of the chief's son." (Arthur C. Ballard. *Mythology of Southern Puget Sound*. University of Washington Publications in Anthropology 3(2):31-150. Seattle. 1929. Pp. 85-86).

The historical reality of mythologies is not something we are arguing here. However, we do understand persistent mythologies such as this one which survived in several versions at least into the first quarter of the nineteenth century are poetic or allegorical representations which probably had at one time specific historical significance, particularly when other ethnographic evidence supports the story. For example, John Peabody Harrington recorded this about the "stakabf" (his orthography) on the White River: "...were friends to the snakes. Snakes were good to them — snakes would not hurt them...that tribe was only one that way" (The Papers of John Peabody Harrington. 1910 Duwamish Field Notes. Microfilm Volume 1 Reel 15 Frame 0496. University of Washington).

3 The Gig Harbor village was called *sqwap‹bc* or *spob‹bc* (Smith 1940: P. 11). The Wollochet village was called *sxwlåtsid*. John Winterhouse, Jr. in his archeological survey of lower Puget Sound recorded several sites at Wollochet Bay (John Winterhouse, Jr. "A Report of an Archaeological Survey on Lower Puget Sound" nd:3-6). On May 20, 1792, Archibald Menzies, the botanist with the Vancouver expedition, observed Indians gathering and drying clams in Wollochet Bay:

"...as we began to pull towards them (the Indians) we observed the women & children scudding into the woods loaded with parcels, but the Men put off from the shore in two Canoes to meet us, we made them some little presents to convince them of our amicable intentions, on which they invited us by signs to land & the only one we found remaining on the Beach was an old woman without either hut or shelter, setting near their baskets of provision & stores, the former consisted chiefly of Clams some of which were dried & smoaked & strung up for the convenience of carrying them about their Necks, but a great number of them were still fresh in the shell which they readily parted with to our people for buttons beads & bits of Copper." (Archibald Menzies. Menzies' Journal of Vancouver's Voyage, April to October 1792. C.F. Newcombe, ed. Archives of British Columbia Memoirs 5. Victoria, B.C. 1923. Pp. 33-34).

The Quartermaster Harbor village was called *tsugw‹lEl*. The Wollochet village sites were also mentioned by Menzies and confirmed to Winterhouse by one of his informants, Puyallup allottee Jerry Meeker. Louisa Duette, in her Court of Claims deposition, recalls seeing the houses at Gig Harbor. (Court of Claims: P. 642). Jerry Meeker told George Chute about the big houses at Gig Harbor (George Roger Chute Collection, Chute (MS 15), Box #4/104. Special Collections, Washington State Historical Society).

4. Smith 1940: P. 11.

5. John Swanton. *The Indian Tribes of Washington, Oregon & Idaho*. Bulletin 145. Bureau of American Ethnography. Smithsonian Institution. Washington, D.C. 1952. P. 37.

6. George Gibbs. *Indian Tribes of Washington Territory*. Reports of Explorations and Surveys, to Ascertain the Most Practical and Economical Route for a Railroad from the Mississippi River to the Pacific Ocean, Made Under Direction of the Secretary of War, in 1853-4, According to Acts of Congress of March 3, 1853, May 31, and August 5, 1854. (Reprinted: Ye Galleon Press, Fairfield, WA, 1967). P. 41.

7. Herbert C. Taylor, Jr. *Anthropological Investigation of the Medicine Creek Tribes Relative to Tribal Identity and Aboriginal Possession of Lands*. Pp. 401-473 in Coast Salish and Western Washington Indians, II. (American Indian Ethnohistory: Indians of the Northwest). Garland Press. New York. 1974. P. 419.

8. Taylor: Pp. 426-427.

9. Lucy Gurand is Lucy Gerand in Waterman. Her name is written at other times as Lucy Slagham and Lucy Schlom. "She is about 85 years old" Court of Claims: P. 646.

10. The newspaper does not provide an age for Lucy Slagham at the time of the testimony, however, an 1888 census of the Puyallup reservation lists her age as forty. If this record is a more accurate accounting of her age than that in the Court of Claims text she would have been around eighty at the time of testimony. She would have been around six years of age at treaty time.

11. Thomas Talbot Waterman worked for the Museum of the American Indian from 1919-1922. He was "employed at the University of Washington" and was paid by the University to conduct an ethnographic survey during the summer of 1918 and 1919. He published a number of monographs of interest to students of Puget Sound cultures including *Notes on the Ethnology of the Indians of Puget Sound*, Indian Notes and Monographs, Miscellaneous Series No. 59, Museum of the American Indian, Heye Foundation. New York: 1973, "The Geographical Names Used by the Indians of the Pacific Coast, The Geographical Review, Volume XII. The American Geographical Society, New York:1922, *Types of Canoes on Puget Sound*, Museum of the American Indian, Heye Foundation. New York: 1920. His extensive list of place names in the Sound is to be found in his Puget Sound Geography, Manuscript

No. 1864. National Anthropological Archives. Smithsonian Institution, Washington, D.C. Microfilm copies of this are available to be reviewed in other library collections.

12. See Marian Wesley Smith Collection. Manuscript 268 (Microfilm) British Columbia Archives and Record Service. Victoria, B.C. Box 8:4:3. Ballard interviews with Joe Young in 1924 place Gurand there then and Janet Haugen has a photograph of her digging clams "circa 1928" in her remembrances of the Kingsbury estate, originally published in the Nor'Wester in 1934. (*Pioneers of Vashon and Much, Much More.* Janet Haugen. 1985).

13. Waterman 1922: P. 180.

14. Waterman 1922: P. 180.

15. Waterman 1922: P. 185.

16. Waterman 1920: Pp. 34-40.

17. Lucy Gurand is remembered fondly by the Kingsbury family of Vashon Island in Janet Haugen's book. According to Haugen, Lucy Slagham Gurand with her husband sold her allotment and remained on Vashon until her death, which would have been not more than a year or two after her testimony in the Court of Claims.

18. Court of Claims: P. 653.

19. Court of Claims: P. 657.

20. Court of Claims: P. 664.

21. According to Benny Mowitch of Taholah whose father and grandfather were Puyallup allotees, the Wollochet area was occupied by Indian people, including his kin, through the 1930's. Dave and Annie Squally, Puyallup allottees, were resident there, along with other kin, according to 1929 Puyallup enrollment documents. Dave and Annie Squally were also "Whollochet Bay Informants" to T.T. Waterman. Waterman in "Puget Sound Geography" drafts notes that Squally was "an excellent informant" and was father to Emma Simmons, one of his informants for place names on Dye's Inlet and Port Washington. Waterman notes that Sally Jackson also lived at "Whollochet Bay" and that Burnt Charlie lived there with the Squallys. Dave Squally died in 1936 and after his death, Annie Squally sold their land in Wollochet and moved to the Puyallup Reservation. (Peninsula Gateway, August 15, 1984). Annie Squally died in 1940 at age eighty-seven after fracturing her hip. In 1878 U.S. Indian Agent Milroy compiled a list of "heads of families and the no. of each, and of individual Indians belonging to the Gig Harbor and Steilacoom, bands of Puyallup tribe of Indians in Pierce Co. Wash. Ter." (Robert Milroy. "List of heads of families and individual Bands of the Puyallup Tribe," May 31, 1878. Milroy Papers. Puyallup Tribe Archives). These are, presumably, people living off reservation and known to Milroy. They number forty-six including seventeen adult men, thirteen women, and sixteen children.

22. John Xot's name appears, for example, on the "List of Indians Who Have Selected Lands on the Puyallup Reservation for Homes," January 25, 1875, in the Puyallup Archives. He is also a contributor to Ballard's *Mythology of Southern Puget Sound.*

23. Arthur Ballard worked primarily in the Auburn area and with people who lived traditionally in the Green River and White River areas. He also collaborated with T.T. Waterman during the summer of 1918. He published several works under his name including "Calendric Terms of the Southern Puget Sound Salish" (*Southwestern Journal of Anthropology*, Volume 6, Number 1, Spring 1950) and *Mythology of Southern Puget Sound* (University of Washington Press, Seattle, 1929). His informants for *Mythology* included John Xot, whom he identifies as "Lower Puyallup. Born about 1845." Other informants included: John Simon "Upper Puyallup. Born about 1840," Tom Milroy "Upper Puyallup. Born about 1845," Dick Swatub "Lower Puyallup. Born about 1840," Joe Young "Puyallup. Born 1863," Charley Ashue "Yakima-Puyallup. Born about 1855," Burnt Charlie "Puyallup. Born about 1835." (Burnt Charlie lived with Dave and Annie Squally at Wollochet during the time that T.T. Waterman was interviewing for place names.) John Xot's maternal grandfather, ko'ialkW, lived in the *sxwobabc* village at Quartermaster Harbor and the father of his maternal grandfather was at Gig Harbor according to Ballard's notes in Smith (Smith Collection: Box 8:4:14). Ballard notes that ko'ialkW's father was also *sxwobabc* (and part Skokomish).

24. Smith Collection: Box 8:4:51.

25. Smith Collection: Box 8:4:3. Waterman notes that the people called *Sxwob-ábc* were the "Swift-water dwellers" (Puget Sound Geography).

26. Smith Collection: Box 8:4:3.

27. Smith Collection: Box 8:4:3.

28. It seems very possible, however, that of these, only Dr. W.F. Tolmie spoke and understood Salish.

Others probably worked in and through Chinook jargon or through translators (cf. Taylor: P. 411).

29. Henry Sicade was a chief elect of the Puyallup Tribe. Some of his recollections were recorded by Elizabeth Shackleford in 1916 and 1917, and he produced a piece on the meaning of "Puyallup" for the Tacoma Evening News (published June of 1916).

30. D.W. Meinig. *The Shaping of America: Continental America 1800-1867.* Yale University Press. New Haven. 1986. P. 119.

31. Norman Graebner. *Empire on the Pacific.* Ronald Press. New York. 1955.

32. Graebner: P. 219.

33. Marian W. Smith, "The Puyallup of Washington," P. 4-5 in *Acculturation in Seven American Indian Tribes.* Ralph Linton, ed. D. Appleton-Century. New York. 1940. (Reprinted: Peter Smith, Gloucester, MA, 1963).

34. Smith 1963: P. 6.

35. See Waterman's *Types of Canoes on Puget Sound* for more detail.

36. T.T. Waterman. "Native Geography in the Puget Sound Region." in Puget Sound Geography.

37. Smith 1940: Pp. 29-30.

38. These berry and root grounds were managed in a number of ways. For a discussion of research see *Indians, Fire and The Land in the Pacific Northwest.* Robert Boyd, ed. Oregon State University Press. Corvallis: 1999. There are several mentions of the use of fire in land management in depositions by Puyallup people in Court of Claims.

39. Smith 1940: P. 32 .

40. For example, Wapato John in his deposition recalls a house over a hundred feet long in the Lime Kiln area (Court of Claims: P. 656).

41. Lucy Gurand, Court of Claims: P. 648-649.

42. Waterman places this at a mouth of stream that ran in a gully near 24th street of the present city of Tacoma. He transcribes it *Tuxwa'dabcEb* said to mean "ground flooded or dry according to the tides."

43. Waterman calls this site *Casqwo'd-tsid.* The father of John Knott (Xot), one of Waterman's and Ballard's informants, was the "principal man here."

44. For a brief discussion of this practice with specific reference to Puyallup predecessors bands see Court of Claims: P. 630 (25 March 1927 Deposition, Jerry Meeker, Puyallup); P. 671 (25 March 1927

Deposition, Joseph Swyell, Puyallup); and P. 677 (25 March 1927 Deposition Mary Anne Dean, Puyallup):

> Question (to Mary Anne Dean): Did they cultivate the wild roots that they used?
>
> Answer: She said when they dig the roots, that softens the ground for the younger sprouts, and the following year they come up again. They sometimes used to burn the grass or burn the underbrush to make it grow better the following year.
>
> Question: Did they plant the little roots back in the soil?
>
> Answer: The seeds go back and then the fine roots grow up, again after it is taken up.
>
> Question: Did they burn the underbrush in the woods so as to keep the trees growing and to keep the fires out?
>
> Answer: Yes sir; that is what they did to make the hunting ground.

45. For a more complete list of plants, their properties and their use by Puget Sound people, the reader may consult Smith's *Puyallup-Nisqually* Chapter VII, *Ethnobotany of Western Washington: The Knowledge and Use of Indigenous Plants by Native Americans,* Erna Gunther, University of Washington Press, and *Plants of the Pacific Northwest* edited by Jim Pojar and Andy MacKinnon with ethnobotanic entries by Nancy Turner.

46. Smith 1940: P. 233.

47. Arthur C. Ballard. "Calendric Terms of Southern Puget Sound Salish." *Southwestern Journal of Anthropology* 6(1):79-99. Albuquerque. 1950. P. 80.

48. Smith 1940: P. 26.

49. Meeker Family Notebooks. University of Washington Library, Manuscripts. Vertical File 94: Notebook I, Pp. 88-89.

50. T.C. Elliot. *Journal of John Work. November and December, 1824.* Washington Historical Quarterly 3 (3) :198-228. Seattle. 1908. P. 212.

51. *The Journal of John Work, January to October, 1835,* with an introduction and notes by Henry Drummond Dee. Printed by C. P. Banfield. Victoria, B.C. 1945. P. 92.

52. Gary Reese. *Nothing Worthy of Note Transpired Today. The Northwest Journals of Augustus V. Kautz.* The Tacoma Public Library. 1978. P. 11.

53. Taylor: P. 427.

54. Joseph Lane. Letter in *Report of the Commis-*

sioner of Indian Affairs. 31st Congress, 2nd Session, 1849-50. Senate Executive Document 1.

55. Taylor: P. 419.

56. W.F.Tolmie, Letter February 17, 1854. Records of the Washington Superintendency of Indian Affairs, 1853-1874. Miscellaneous Letters Received August 22, 1853-April 9, 1861. Microfilm No. 5, Roll 23.

57. George Gibbs. *Indian Tribes of Washington Territory.* P. 41.

58. George Gibbs "Census of Indians in Western Division of the Territory. February 1854." These are among miscellaneous notes and drafts for Gibbs' *Indian Tribes of Washington Territory.* Bureau of American Ethnography. Wash. Ethnol. 2356.

59. Map of Washington Territory Showing the Indian Nations and Tribes. Copied from the original preliminary sketch compiled from the archives of the Pacific Rail Road Expedition and Survey under the direction of Gov. I I. Stevens. J. Lambert topographer of the expedition. Indian names and boundaries by G. Gibbs, 1853-54. San Francisco California April 14, 1854. Copy in Tacoma Public Library, Tacoma, Washington.

60. Map of Washington Territory Showing the Indian Nations and Tribes. By permission of Gov. I.I. Stevens. J. Lambert, Draughtsman. Traced from Mr. Lambert's original map by Geo. W. Stevens. Washington State Historical Society Archives. Tacoma, Washington.

61. Washington Territory West of the Cascade Mountains showing the Boundaries of lands ceded at Treaty of Dec. 26th 1854. Original in National Archives and Records Administration. Washington, D.C.

62. Washington Territory west of the cascades showing the boundaries of lands ceded at treaty of Dec. 26, 1854 and the reserves. Also Indian tribes to be treated with, and lands to be ceded at future treaties. Original in National Archives and Records Administration. Washington, D.C.

63. George Gibbs. Cascade Road – Indian Notes. No. II. 1854-55. National Archives. Washington, D.C. RG 76. International Boundaries, Northwest Boundary-Summit of Rocky Mountains to Pacific Ocean. Box 3.

64. I.I. Stevens' letter to Col. M.T. Simmons, March 22, 1854. Records of the Bureau of Indian Affairs. Oregon Superintendency Field Papers. Microfilm. Washington State Library, Olympia, Washington. Copy in ICC Docket 296 "Exhibits."

65. Map of the Western District of the Washington Territory Showing the Position of the Indian Tribes and the Lands Ceded by Treaty. Drawn by George Gibbs 1855. Copy Washington State Library, Olympia, Washington.

66. George Gibbs. Report to Gov. I.I. Stevens. Olympia. 1856. "Introduction to a Report to the Hon. I.I. Stevens, Governor and Superintendent Indian Affairs of Washington Territory, on the Indian Tribes of the Western District." (corrected handwritten manuscript) Bureau of American Ethnography. No. 743.

67. George Gibbs. *Tribes of Western Washington and Northwestern Oregon.* Contributions to American Ethnology 1(2):157-361. John Wesley Powell, ed. U.S. Geographical and Geological Survey of the Rocky Mountain Region. Washington, D.C. 1877. P. 178

68. Letter from Franz Boas to his wife. Victoria, B.C. July 30th, 1890. American Philosophical Society. He repeated this news in a letter to his parents. Letter from Franz Boas to his parents. Westmister, B.C. July 31st, 1890. American Philosophical Society. He wrote to his wife on July 27 from Tacoma where he was staying in The Tacoma hotel while working at the Puyallup Reservation. He wrote that the agent had sprained his back and was in bed and that he had been compelled to go to church by the agent's wife and his daughters. He was fearful that as an "honored guest" he would have to offer a prayer and complained about the whole church experience. We have not yet located his field notes from the Puyallup study.

69. Edwin Chalcraft. *Memory's Storehouse.* Unpublished memoir. nd. P. 83.

70. John Winterhouse, Jr. *A Report of An Archaeological Survey on Lower Puget Sound.* Unpublished. nd. Meeker.

71. Melville Jacobs Collection. University of Washington Archives, Box 120-34, #14773.

72. Including John Xot, Joe Young and others noted above.

73. Hermann K. Haeberlin and Erna Gunther. *The Indians of Puget Sound.*

4

Field and Laboratory Methods and Procedures

Mary Parr
Julie K. Stein
Laura S. Phillips

This chapter includes the step-by-step methods and procedures used by the investigators of the Burton Acres Shell Midden. To accommodate public participation, our field and laboratory methods differ from those traditionally used in archaeology. Therefore, we provide detailed explanations of our methods to assist archaeologists who wish to undertake similar projects. This public education project was designed so that each volunteer would experience the entire field and laboratory process rather than just unearthing artifacts. Only in this way would people begin to understand the nature of archaeological fieldwork and the important fact that digging is only a fraction of the entire process.

The field and laboratory methods used at the Burton Acres Shell Midden were modeled after the protocol developed for the British Camp, also known as English Camp (45SJ24), excavation directed by Dr. Julie K. Stein (Stein et al. 1992). British Camp was excavated by field school students, and the protocol was designed for people who knew about archaeology from classroom preparation but who had no previous field experience.

The Burton Acres Archaeological Project was unique in that untrained members of the general public were the excavators. Interestingly, few modifications to the protocol were necessary to accommodate this difference. Modifications were made to address on-the-spot training and add checks to catch and eliminate errors

To facilitate the goal that each volunteer experience the entire field and laboratory process, Burton Acres County Park was divided into five distinct stations: Check-in, Excavation, Screening, Sorting, and Identification/Cataloging (Figure 4.1). These stations were

indicated by yellow and white awning tents. The tents were useful in delineating the various stations and in keeping the participants dry during the occasional rainy weather. The brightly-striped awnings and the nature of the project generated feelings of excitement and interest in the whole process.

In this document each of the five stations will be explained, as well as the means and methods by which the archaeological material was processed at each station.

CHECK-IN

Over 375 members of the general public participated in the project as volunteer archaeologists. The volunteers were scheduled in advance and notified in writing of their start times. They were instructed to arrive fifteen minutes early for the Check-in process.

At the Check-in table, the volunteers were greeted by a staff archaeologist (Figure 4.2). Volunteers were supplied with a clipboard containing a set of three color-

Chapter opening photo: Mike Etnier, a University of Washington archaeologist, teaches an eager volunteer how to identify shells, animal bones, metal, and charred botanical remains.

Figure 4.1 The stations are identified by yellow and white striped awnings: Check in, Sorting, and Identification/Cataloging (right), Screening (middle), and Excavation (distant left).

Figure 4.2 The Check-in area was situated to meet visitors as they left the parking lot and immediately behind the park regulation signs.

coded forms. The forms became the official archive record of the project and thus were treated with great care. They were color-coded for easy recognition and consisted of a yellow <u>Bucket Form</u>, a blue <u>Screen / ID Form</u>, and a two-page pink <u>Identification Form</u>. Each clipboard also held four bag tags designed to accompany the excavated material (these were eventually placed in bags) through the screening and sorting process (see Appendix B).

At Check-in, the volunteer was instructed to write his or her name, the date, and the start time on each form (Figure 4.3). After this was completed, volunteers were given a "dig kit" consisting of a garden-tool caddy containing a trowel, a dustpan and broom set, a metric folding rule, and a pencil. From Check-in, volunteers were escorted to the Excavation Station.

EXCAVATION

The Excavation Station was located furthest from the parking lot and required that people walk out to the most easterly point of land. Four (1m by 1m) excavation units were opened. One awning was placed over the two most northerly excavation units, because they were adjacent to each other and could take better advantage of the awning's protective cover in the sun or rain. The other two units were uncovered (Figure 4.4).

Units, Layers, and Depth

Each excavation unit was 1m by 1m in horizontal dimensions, and named with the grid coordinates of the

entire unit (Figure 4.6). The elevation of the surface was measured from Datum A (Figure 4.5). The exact elevations of each corner for the four excavation units were recorded on stakes defining the corners and on the <u>Unit/Layer Form</u>. To measure the depth below surface, the rod was stretched across the unit over the spot to be measured. A folding metric rule was used to determine the distance from the rod down to the object/point. This metal rod became known as "The Rod of Science" (a term coined on *Bill Nye the Science Guy*, a television program popular at the time). The placement of the rod (north/south or east/west) did not change the depth below surface at the point more than +/- 1cm (Figure 4.7). This error was acceptable given the convenience of the procedure.

Excavating

When each volunteer arrived at the Excavation Station, he or she was directed to a table where staff archaeologists instructed them as to which unit, layer, and bucket number they would be excavating. The volunteer completed that information on the forms before proceeding to the excavation units. If the volunteer was under twelve years of age he or she was also accompanied by an adult family member. The volunteer also completed four bag tags with the appropriate unit and layer information, one for each bag of the following four screen sizes: 1 inch (25.4 mm), 1/2 inch (12.7 mm), 1/4 inch (6.35 mm), and 1/8 inch (3.175 mm). These bag tags were placed in the bottom of the bucket bag and

Figure 4.3 Tegan Horan and Sage Alderson-Gamble assist volunteer Patricia Baillargeon in filling out forms during check-in.

Figure 4.4 Archaeologists (in white shirts) teach volunteers how to excavate. Note in the right background a student giving a tour to visitors.

remained with the excavated material throughout the entire process. Names of volunteers and their respective bucket numbers were recorded on the <u>Unit/Layer Form - Additional Comments - Bucket Tally Form</u> with a separate list for each unit (Appendix B).

The staff archaeologist then escorted the volunteer to one of the four excavations units and marked an area of approximately 25cm by 25cm within the 1m by 1m unit. The volunteer used a trowel to carefully remove the sediments from this 25cm by 25cm area, to a depth of approximately 2cm to 3cm. The volume resulting from this excavation filled a small bucket with 2 liters of site material. Not all the excavation areas were exactly 25cm by 25cm. Some were 15cm by 35cm, or any other shape that was useful for excavating 2 liters of material. Each volunteer was given a trowel and dustpan, and then instructed on troweling a shell midden. The volunteers excavated much more slowly than the professional archaeologists. Some expressed fear that they would damage something, others were just enthralled with every little shell or rock. In the end, most had to be encouraged to complete the troweling portion of their task.

The 25cm by 25cm area was then mapped on the <u>Bucket Form</u>. The outline of the area was drawn by the volunteer on a 1m by 1m plan view grid with 10cm gridlines. The point in the center of the area was designated as the provenience point. The north-axis, east-axis, and depth below surface were recorded for this provenience point, which represented the bottom of the excavation area and corresponded to the bucket number. Thus, provenience information was recorded for every 2-liter bucket. This mapping technique proved valuable when describing the stratigraphic layers in each unit, and when selecting sediment samples to analyze.

At the end of each day's excavation, Stein took elevations of the surface of every 1m by 1m excavation unit and wrote notes on the <u>Unit/Layer Form - Additional Comments Form</u> for each unit. These notes recorded the progress made and also instructed the next day's staff archaeologists. All excavation <u>Unit/Layer Forms</u> were kept in notebooks at the Excavation Station, and were brought back to the laboratory every night.

Cataloging at Excavation

The bucket was the basic unit of provenience for the excavation. All excavation material could be traced to a 25cm by 25cm location, with provenience of unit, layer, and bucket number (e.g., 22582A015 was from Unit 22/58, Layer 2A, Bucket Number 015). All material from the bucket was mapped, screened, sorted, and cataloged together.

Occasionally objects were found that warranted extraction as field specimens, with more detailed provenience than just that of unit, layer, and bucket. These objects were listed individually on the <u>Bucket Form</u> and mapped using north-axis, east-axis, and depth below surface. Field specimens were placed in polypropylene plastic zip lock bags, labeled with an acid-free

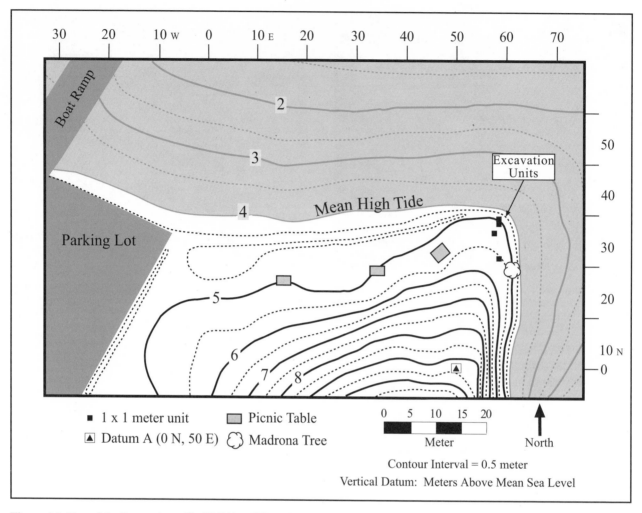

Figure 4.5 Map of the Burton Acres Shell Midden differentiates the shaded area that is inundated at high tide from the white area that is dry land covered in grass. The excavation units are all at the eastern edge of the grass, near the Madrona tree.

label, and sent with the volunteer to be cataloged with the bucket material. Samples taken for radiocarbon analysis were first wrapped in aluminum foil and then placed in a zip lock bag. Very few objects were given field specimen status, as the provenience of each bucket was already at such a fine scale.

Sampling

As is common on numerous excavations of shell middens in the Northwest, the project used a sampling strategy for the smallest material recovered (screening procedures divided each bucket into fractions of various sizes: 1 inch, 1/2 inch, 1/4 inch, and 1/8 inch). Based on the distribution of sizes and richness of material at the Burton Acres Shell Midden site, all material except fire cracked rock from the 1 inch and 1/2 inch screen fraction was saved. In addition, 100% of the bone, lithics, historic artifacts, and other modified tools were recov-

ered from the 1/4 inch and 1/8 inch fraction. Twenty-five percent of the shell was sampled from the ¼ inch and 1/8 inch screens. Charcoal was collected from the 1/4 inch fraction, but not the 1/8 inch screens.

To specify which buckets would be sorted in their entirety, and which would be partially sorted and returned to the site, the staff archaeologists made sure that the volunteer noted at the very bottom of the <u>Bucket Form</u>, either RETURN or SAMPLE. Staff archaeologists designated a bucket as RETURN or SAMPLE by looking at the number of the bucket. If the bucket number was either bucket #1 or a number that is a multiple of four (e.g., 4, 8, 12, 16, etc.), the volunteer was instructed to circle SAMPLE. If the bucket number was not number one or a multiple of four, the volunteer was instructed to circle RETURN.

Buckets labeled as RETURN were examined at the

Figure 4.6 Laura Phillips (left) holds a ruler while a volunteer notes on his form the depth of the bucket he just excavated.

Sorting Station and the shell, 1/8 inch charcoal and rock were not saved for further analysis. This shell, charcoal and rock was eventually returned to the excavation holes.

Buckets labeled as SAMPLE were sorted in their entirety and all objects except the mineral fraction were saved for further analysis.

The excavation process took about thirty to forty-five minutes to accomplish. Volunteers then took their bag of excavated material (each bucket was lined with a plastic bag), and were sent to the Screening Station to begin the analysis of the material.

SCREENING

The Screening Station consisted of two sets of nested screens (Figure 4.8). Each bucket excavated was dry-screened, which proved to be satisfactory for removing most of the matrix. The screens were wood-framed with dowel handles and an outside dimension of 24 inches long by 20.75 inches wide by 4.5 inches deep. They were built by AEOSCREEN of Silver City, Nevada, to fit existing metal frames. The screen mesh sizes were 1 inch, 1/2 inch, 1/4 inch, and 1/8 inch.

When the volunteer arrived at the Screening Station, the staff archaeologist checked to make sure that all forms had been filled out correctly. The total weight of the bucket was measured using a hanging metric scale accurate to +/- 25 grams. The total weight of the bucket was recorded on the bottom of the yellow Bucket Form.

Figure 4.7 Archaeologist Mike Etnier helps a volunteer place a ceramic sherd in a field specimen bag. The rod that appears under Mike's legs in this photo is known as the "Rod of Science" and is placed over the excavation unit to facilitate measuring the sherd's depth.

The volunteer poured the material from the bucket bag into the top screen (1 inch) and shook it to remove all particles smaller than the opening of the screen (Figure 4.9). Each nested screen was sequentially shaken, and then separately moved to a nearby table where the material was removed and placed in a bag with the appropriate bag tag. Each bag of 1 inch, 1/2 inch, 1/4 inch and 1/8 inch material was weighed and the amounts entered on the Bucket Form. The volunteer had now transformed their bucket bag into four new bags. All four bags were placed in the original bucket bag and the volunteer was sent on to the Sorting Station. This activity at the Screening Station took only 20 minutes to complete.

SORTING

The Sorting Station was the largest station because the sorting process took the most time. The sorting area consisted of an awning over four tables and seating for about fourteen people (Figure 4.10). The volunteers

Figure 4.8 Volunteers Hugh and Ellen Ferguson (left) pour the contents of their bucket through the nest of screens. Miranda Stockett (right) is the archaeologist assisting them in the process.

Figure 4.9 A volunteer examines the contents of the 1/2 inch screen after shaking. People bagged objects together to speed up screening and allow identification at the next area (Sorting).

were taught how to identify tiny bits of shell and bone, and how best to systematically separate the mixture on the tray before them. Unexpectedly, much of the education occurred at this station. Some volunteers were very nervous at the excavation area and their anxiety prevented them from asking questions or hearing answers. They were entirely focused on their task. When they finally sat down at the sorting table and realized that they could complete the task in front of them, they began to ask questions and to interact with each other. A tremendous amount of information and understanding (from King Tut to repatriation) was transmitted and received during this activity.

Each volunteer was given an orange food tray (acquired at a restaurant supply store and resembling the trays one gets at a fast-food restaurant) and instructed to sort the material they recovered from their 2-liter bucket (Figure 4.11). They were first directed to pour out the material recovered from the 1 inch screen, and sort it into piles defined by material type. Some materials were saved and some returned to the site, depending on the material, the size fraction, and whether the sample was designated as a SAMPLE or RETURN bucket (see previous discussion on sampling). Each size fraction was sorted completely, before sorting began on the next size fraction.

Material Types

The material types for the project were defined as: angular and rounded rock, plant, shell, bone, lithic,

charcoal, glass, metal, ceramic, basketry, and other historic material. These objects were saved in the following fractions and samples:

Angular and Rounded Rock: The 1 inch and 1/2 inch rock were divided into two groups each: rocks with angular edges and those with rounded edges (Figure 4.12). Angular and rounded rock were weighed separately, and placed in a container of materials to be returned to the excavation holes at the end of the project. The weight and count of the angular and rounded rock were recorded on the Screen/ID Form. All 1/4 inch and 1/8 inch rock was returned without being sorted.

Plants: Seeds and wood (charred and uncharred) were saved from all fractions, except the 1/8 inch fraction.

Shell: All shell was saved from the 1 inch and 1/2 inch fractions. All shell was saved from the 1/4 inch and 1/8 inch fractions of SAMPLE buckets. Shell in the 1/4 inch and 1/8 inch fractions of RETURN buckets was returned to the site.

Bone: All bone (mammal, fish, and bird) was saved from every size fraction and every sample.

Lithics: All lithics were saved from every size fraction and every sample.

Glass: All glass was saved from every size fraction and every sample.

Figure 4.10 Volunteers sit at tables in the sorting area while archaeologists lean over their shoulders teaching them to identify shell from bone and rocks.

Figure 4.11 Two volunteers spread the contents of the their 2-liter bucket onto a sorting tray. The woman is asking a nearby archaeologist if their identification of an object is correct.

Metal: All metal (mostly corroded bits identified by an oxidized reddish color) was saved from every size fraction and every sample.

Ceramic: All ceramics were saved from every size fraction and every sample.

Basketry: All basketry (string and twine) was saved from every size fraction and every sample.

Other Historic Material: All other historic material (such as plastics and fiberglass) was saved from every size fraction and every sample.

In the original protocol the 1/4 inch and 1/8 inch RETURN buckets would be examined only for significantly modified objects, since a 25% sample of the 1/4 inch and 1/8 inch fauna and flora would be collected in the SAMPLE buckets. This plan, however, was altered in the first few days of sorting. The controversy began when the staff archaeologists noted that what was significant to the volunteer included more categories of objects than what the staff considered significant. The staff considered such things as beads, bone points, and lithics as significant. The volunteer considered seeds, charcoal, bones, glass, metal, and plastic as significant.

The protocol was changed to include all 1/4 inch and 1/8 inch cultural material from the RETURN buckets except shell, but in these first days some seeds, charcoal, bones, glass, metal, and plastic were probably placed in the RETURN container, along with the shell and all the rocks. The net result was that a nearly 100%

sample of every class of artifact except shell was collected during this project. One hundred percent of the 1 inch and 1/2 inch shell, and 25% of the 1/4 inch and 1/8 inch shell, was collected.

Data Recorded

At the Sorting Station, the volunteer was encouraged to let the staff archaeologists fill out the <u>Screen/ID Form</u>. Some people wanted to experience everything and were allowed to fill in the blanks, but too many errors were observed (and had to be corrected) and thus the practice was discouraged. The <u>Screen/ID Form</u> had predetermined spaces for the weights and counts of every material type. The volunteer was asked to call out the counts and weights of each material type, and the staff entered the data on the form.

One of the most difficult tasks at the Sorting Station was to teach the volunteer the difference between rocks with rounded edges and angular edges, and why the distinction is important. The volunteer had to consider the last thing that happened to the rock. Was it smoothed by erosion or was it broken by some force such as cracking in a fire or falling off a cliff? If a rock had any edge that was angular it was classified angular, even if most edges were rounded (Latas 1992). This distinction was made because the fractures probably occurred as a result of cultural practices (stresses induced by heating and cooling, or by pounding) and therefore are artifactual in nature. Once this separation was made, the angular-edged and rounded-edged rocks

Figure 4.12 This young volunteer asks if the object he is holding is a rounded rock. He was right.

were counted and weighed. These two numbers were recorded on the <u>Screen/ID Form</u>. This separation was made for only the 1 inch and 1/2 inch rock.

All material was weighed on two electronic balances, a Shimadzu Libror EB-5000 and an Ohaus GT-2100. One balance was accurate to the nearest tenth and the other to the nearest hundredth of a gram. The most accurate balance was used except in extremely busy times, when people needed to be processed quickly to insure the safe-keeping of all the bags and forms.

The total weight and count of bones was recorded by staff archaeologists on the <u>Screen/ID Form</u>, noting the condition (burned or not burned). The bones were then separated by staff into fish, mammal, bird, and unidentified. The weights and counts were recorded for each. The bones were bagged separately in small zip lock bags with their faunal identification and screen size written on the bag.

This recording process was repeated for all material types and all screen sizes, noting the total weight, count, and the condition. Shells were separated by clam, mussel, or other and bagged and recorded separately. Lithics were separated by chert, red chert, dacite, obsidian, petrified wood, or other. They were then counted and weighed. Historic materials were also weighed and recorded on the <u>Screen/ID Form</u>. A code was devised for the historic materials (e.g., G=glass, M=metal, C=ceramic, S=non-charred seed).

Occasionally, an uncommon material type turned up in the excavation. For example, several fragments of

plastic and a length of rubber hose were recovered. At the beginning of the project, these items were given codes like R for rubber and P for plastic. This quickly led to code F for fiberglass and N for nylon fishing line. To simplify the cataloging process, objects that fell outside the previously mentioned material types (glass, metal, ceramic, or seed) were all given the code H for "historic other."

Buckets that were designated as SAMPLE took much longer to sort. All shell had to be separated from rock, which took most people almost five hours as opposed to one or two hours for the RETURN samples. The shell from the 1/4 inch and 1/8 inch fraction was too numerous to count in the field, so an estimated count was made and recorded on the <u>Screen/ID Form</u> as 20+, 50+, 100+, etc. The 1/4 inch and 1/8 inch shell was not sorted by species in the field. This identification was done by the analyst (see Chapter 11).

For categories in which no objects were found, NONE was written by the staff archaeologist in the appropriate boxes. For buckets that were not bucket #1 or a multiple of four, RETURN was written on the <u>Screen/ID Form</u>. This was helpful every evening during inventory/catalog check to determine if a bag was missing or just did not exist.

CATALOG/IDENTIFICATION

When all four screen sizes were sorted, weighed, counted, and bagged, the volunteer proceeded to the Catalog/Identification Station. The Burton Acres Archaeological Project was designed to illustrate to each individual all the steps involved in excavation, including the identification and cataloging that usually occurs months later in a lab. Thus, this station was designed as the place to identify the artifacts in the field. It was also a chance for the volunteer to see how his or her artifacts related to others, and to see the emerging information about the site.

During the identification process, the staff archaeologists had an opportunity to discuss the importance of provenience and context, fundamental concepts in archaeology, with the volunteer and to emphasize preservation and cultural resource management laws. Provenience is the three-dimensional spatial position of an object. Archaeological context is the recording of stone, bone, metal and all other objects to emphasize their associations, matrix, and provenience.

The volunteer came to the Catalog Station with a clipboard full of all the forms, plus the plastic bags from the sorting process, and any plastic bags of field specimens recorded during excavation. The staff archaeologist began by checking all forms for accuracy and completion, including the name, date, unit, layer, and bucket number. The two-page, pink Identification Form was completed while the volunteer watched and listened. The field specimens were weighed and coded after they were described, and the remaining bags were arranged by the staff archaeologist on the table in order of size fraction and material type, and assigned codes.

Field Codes

Codes (abbreviations for material types and locations where objects are found) were used to catalog objects. The codes (Table 4.1) are two-digit unique numbers, one for material type and the other for location. For example, a bone recovered from the 1/2 inch screen would be given the code of 22, 2 for bone and 2 for the 1/2 inch screen. These codes are similar to those used at the excavations of British Camp on San Juan Island, Washington (Stein et al. 1992).

In addition to the material codes, some objects were found in such large numbers that the codes were pre-printed on the field forms (see Appendix B). Those materials were defined as follows.

Bone

Bone was processed by giving the unmodified bones a material code (2) and location code dependent on the screen size from which it came. The bone was weighed and counted, and the general taxon (mammal, bird, fish or unidentified) was assigned. Also noted was whether the bone was whole or fragmentary, cut or not cut, burned or not burned.

Weight measurements were taken using either the electronic balances or with a scale known as the Micro-Evolution scale. This was a small mechanical balance made by Ishida.

Modified Bone Antler

Any modified bone or antler was examined, recording the field specimen number, weight, count, object name, taxon, element, and manufacturing technique.

Lithic

Data about the lithics were recorded on the Identification Form, including weight, count, object name, material type (chert, red chert, dacite, obsidian, petrified wood, other), and manufacturing technique (chipped or

Table 4.1 Material and Location Catalog Code

Material		Location	
Type	Code	Type	Code
plant	0	in situ	0
shell	1	1" screen	1
bone	2	1/2" screen	2
lithic	3	no provenience	3
charcoal	4	1/4" screen	4
glass	5	1/8" screen	8
metal	6	bulk sample	9
ceramic	7		
basketry	8		
combination	9		
historic other	H		

ground). If a lithic turned out to be an unmodified rock, the staff archaeologist discarded it, making a note on the Screen/ID Form that the "lithic" was not a lithic and had been discarded.

Shell

Unmodified shell was examined and identified to family name if possible. Clams were further identified as Butter, Littleneck, Horse, etc. It was noted whether the shell was a hinge (H) or a valve part (V), the percentage present noted either as fragment (F) or whole (W). Also noted was whether the shell was burned (B) or not burned (NB). Other shells, such as mussel, limpet, oyster, etc. were recorded in the same way. For 1/4 inch and 1/8 inch fraction, the counts of shell were estimated, and the shell was not sorted. The total weight and count were transcribed from the Screen/ID Form to the Identification Form.

Historic

For historic objects, the weight, count, name, material type, and the technique of manufacture were recorded.

Modified Other

This category was used for recording the charcoal that was found. It was also designed to record such things as modified shell, or wooden stakes. Both the charcoal found during sorting the 1 inch, 1/2 inch, and 1/4 inch screened material and charcoal taken as field specimens for radiocarbon dating were recorded here. Charcoal was not saved from the 1/8 inch size fraction. If no artifacts were found, the staff wrote NONE in the

Table 4.2 Example of a catalog number dissected to identify meaning of each segment

Example Catalog Number: 29582A0035400	
2958	Unit
2A	Layer
003	Bucket Number
5	Material Type
4	Screen Size
00	Field Specimen Number

appropriate boxes on the pink form. This proved to be very helpful in making sure that everything sorted was also checked and cataloged.

After the field codes were filled out, the volunteer was provided with a catalog number for each object. The catalog number is a combination of the provenience, size, and material of the object (Table 4.2). The unique number is permanently associated with the object, preserving each object's important contextual information. The catalog numbering system for the Burton Acres Shell Midden was first used at British Camp.

LABELING

Labeling was the volunteer's last step in processing the archaeological material. The staff archaeologists wrote the catalog number on a catalog label (Appendix B). The catalog number was the unique thirteen-digit number that recorded the unit, layer, bucket number, material code, location code, and field specimen number of each object or groups of objects. The excavation date and a brief description of the contents of the bag were also written on the catalog label. The self-adhesive acid-free, foil-backed label was then placed on a new appropriately sized zip lock bag, and the contents placed inside.

Each material type and size fraction was placed in a separate new plastic bag with a unique catalog number. For example, all 1 inch unmodified bone was placed together in one plastic bag. If the bones had been separated by animal taxon (such as fish or bird), and were already in separate bags, the smaller bags were placed together in one larger bag with one label and one catalog number.

The last two digits of the catalog number were reserved for a field specimen number. Objects recovered in situ at excavation (radiocarbon samples and some modified objects) received field specimen numbers. On all bags of non-field specimen material, the last two digits were 00.

Modified objects that were identified at the Screening Station or during sorting also received field specimen numbers. The rationale behind this was that a modified object such as a bone awl would be removed from the bag of unmodified bones and placed in its own bag. This object then required its own unique catalog number to maintain its provenience information and provide a unique number for the bag.

After cataloging and labeling, all cataloged objects were placed in acid-free curation boxes in order by material type, unit, layer, and bucket number. The forms were separated and placed in notebooks arranged by excavation unit. The collection and archives are in the possession of the Puyallup Tribe of Indians (2002 East 28th Street, Tacoma, Washington, 98404). Copies of the archives are available in the Archaeology Collections at the Burke Museum.

CHECK-OUT

Lastly, the volunteer was sent to the Check-out Station where his or her empty clipboard was returned. Check-out was located next to Check-in so that the volunteer experienced a sense of completion. A glass-covered specimen box was located at Check-out. It held a selection of the artifacts (e.g., vial of herring vertebrae, a bone point, a scallop shell) found during the project, and was used to educate visitors as well as to give the volunteer a sense of their contribution to the entire project. This display was changed as new objects of interest were found. Volunteers were encouraged to form opinions as to what the finds might mean in the interpretation of the site.

Also at Check-out pamphlets were available with additional information about archaeology. The Puyallup Tribe of Indians provided a three-page history of their people. The State Office of Archaeology and Historic Preservation provided summaries of the laws applicable to historic preservation. A pamphlet produced by various state and federal agencies about the archaeology of Washington was available. In addition, information concerning Burke Museum membership was there. The permit from the State Office of Archaeology and Historic Preservation granting permission to excavate

Figure 4.13 Laura Phillips (left) and Laura Andrew (right) excavate the deepest layers of Unit 22/58 using the larger 8-liter bucket.

Figure 4.14 Student Tegan Horan (left) and archaeologists (from left to right) Laura Phillips, Angela Linse, and Julie Stein work into the evening to complete the excavation of Unit 22/58.

was also posted at Check-out. Volunteers and visitors took advantage of the resources, asked questions, and went away with a greater understanding of archaeology and the local history.

STAFF CHECK-IN

At the end of each day, all artifacts and forms were inventoried and organized by staff archaeologists at the lab provided by the McMurray Middle School. This procedure provided further assurance against information loss and error. All forms were reconciled for each bucket and cataloged artifacts to insure that catalog numbers and provenience information were the same.

In the lab, the bags with the artifacts were arranged in acid-free boxes. Their order was arranged by material type, unit, layer, and bucket number. The attempt of the Burton Acres Archaeological Project was to skip the field lab step and go directly from field cataloging to final curation. Although all material was analyzed later by separate analysts, the cataloging and storage system saved the usual step of transferring from field bags to curation-quality bags.

AFTER THE PUBLIC LEFT

Not all of the site was excavated by volunteers using 2-liter buckets. Some of the site was excavated by staff archaeologists and trained students acting as guides. In these cases both 2-liter and 8-liter buckets were used (Figure 4.13). All buckets, no matter who excavated them, were excavated, screened, and sorted following approximately the same procedure as the volunteers.

The students and staff excavated portions of the site because the bottom of the cultural material had to be reached within the allotted time for field work. Our permit was for 12 days of excavation, and it was clear, after the first few days, that the pace of excavation using only volunteer excavators was not sufficiently fast to finish.

Eight-Liter Buckets

On the final day of the planned public excavation, July 2, 1996, the last volunteer left around 2:00 PM, but not all the units had their cultural material completely removed. The middle unit, Unit 26/57, had been finished on July 1, 1996, and the southern-most unit, Unit 22/58, was finished early on July 2, 1996. The northernmost units, 28/58 and 29/58, however, were not yet finished. Once the last volunteers left, staff archaeologists used 8-liter buckets to excavate, instead of 2-liter buckets. The larger bucket changed a variety of things: the shell midden could be removed faster, fewer forms had to be filled out, a larger area could be excavated (an area 50cm by 50cm) per bucket to a greater depth (5cm). This change sped up the process and allowed the staff archaeologists to remove the remaining cultural material safely, yet expeditiously (Figure 4.14).

By 9:00 PM that night, staff archaeologists reached Layer 3A (sterile) in Unit 29/58. This unit proved to contain larger proportions of whole shell and bone than found in other units, as well as more artifacts. Several antler adze handles were found, as well as several stone and bone tools. Shell and charcoal samples were

Figure 4.15 Sage Alderson-Gamble leans over the west profile of Unit 29/58 pointing to the dark, wet layer at the base of the excavation. Each day, incoming tides dampen this layer, and the layer dries during low tides.

collected. Excavations proceeded to about one meter below surface, a surface that is close to the elevation of high tide. As the depth of the excavation approached the elevation of the high tide, the layer got wetter and wetter. This condition affected the preservation of the material in these layers. For example, two wooden stakes were found preserved in the lower levels of Layer 2. This wet layer appeared darker and was visible at the bottom of the east profile of Unit 29/58 (Figure 4.15).

Wet-Screening

Because the material from Unit 29/58 was wet, it would not fall through the dry screens and had to be wet-screened. Wet-screening is a technique in archaeology where water is used to wash material through mesh openings, as opposed to shaking excavated material in a screen. The 8-liter buckets were taken to the Screening Station and first dry-screened in the way described for the volunteers; weighing and recording the total bucket weight and the weights of the four screen sizes. Because the personnel at the screens were all staff or trained students the larger 8-liter buckets could be further

processed at the screening area rather than being bagged and sent to the sorting area. The angular-edged and rounded-edged rock from the 1-inch and 1/2-inch size fractions were counted, weighed, recorded, and discarded at the Screening Station using the hanging scale (a different scale from the one used at the Sorting Station, and accurate only to 25g). These measurements are less accurate than those taken with the electronic balance for rocks from 2-liter buckets. This procedural change, however, reduced the volume of material to be sorted back at the Burke Museum. The material caught on the screens was wet, however, and would need to be cleaned further. It was bagged and taken to McMurray Middle School where it was wet-screened by Laura Phillips, Steve Denton, and Mike Etnier.

The following day, July 3rd, three staff returned to Unit 28/58 to complete the excavation. Because this unit was up-slope and contained a thinner layer of cultural material than Unit 29/58, the deposits were not moist and could be dry-screened in the usual way, and packed in curation boxes to be taken back to the Burke Museum for sorting.

The 8-liter buckets were handled in a manner similar to the 2-liter buckets. Each bucket had a set of forms that recorded the exact location of the bucket. A list of the bucket numbers and excavators was kept on Additional Comments Forms. Because the buckets were so large and the time so short, the time-consuming process of sorting and cataloging this material was saved for later in the museum. Approximately 35 boxes of unsorted material were taken from the site to be processed in the archaeology lab at the Burke Museum.

Column Samples

After the units were fully excavated, Dr. Stein took column samples for sediment and botanical analyses. These samples were taken from the northeast corners of Units 22/58, 26/57, and 29/58. Samples were taken by excavating a 25cm by 25cm area extended to the north from the profile. Each 5cm increment of depth was bagged separately. Each bag was labeled as a column sample, and with the depth below surface. Samples continued, from the sod (Layer 2A) to the sterile glacial material (Layer 3A). Rocks and sediment were analyzed from these bags, as well as botanical materials. Once all cultural material had been excavated from the units, drawings of the unit walls (called profiles) were made on graph paper. The units were then filled with the material marked RETURN (see

Sampling), and covered with sod.

AT THE BURKE MUSEUM

All cultural material and associated documents were taken to the Burke Museum for final processing and analyses. The wet material from Unit 29/58 and column samples was dried. Unsorted material from Units 28/58 and 29/58 was sorted by volunteers, students, and staff at the Burke, and was completed in December 1996. Analyses began in 1997.

REFERENCES

Latas, T.J.
  1992   An Analysis of Fire-Cracked Rocks: A Sedimentological Approach. In *Deciphering a Shell Midden*, edited by J.K. Stein, pp. 211-237. Academic Press, San Diego.

Stein, J.K., K.D. Kornbacher, and J.L. Tyler
  1992   British Camp Shell Midden Stratigraphy. In *Deciphering a Shell Midden*, edited by J.K. Stein, pp. 95-134. Academic Press, San Diego.

5

Stratigraphy and Dating

Julie K. Stein

When a person first sees an archaeological excavation, they often don't notice the layers exposed in the profiles. These layers, called strata, contain information to tell time and origin. Initially, the archaeologist describes each layer as they see it in the field and collects separately the artifacts from each layer. Subsequently, they interpret the activity that produced each layer and the processes that altered each. What a visitor thinks is only dirt is really the ultimate source of clues to the past.

In technical terms the study of the layers of dirt that make up an archaeological site falls within an established subdiscipline in the geosciences called stratigraphy. *Stratigraphy* is the science dealing with the description of all rock bodies forming the Earth's crust — sedimentary, igneous, and metamorphic — and their organization into distinctive, useful, mappable units based on their inherent properties or attributes (Salvador 1994:137). Archaeologists have borrowed the principles of stratigraphy and applied them to more recent layers discovered in archaeological sites (Stein 1987, 1990, 1996).

The stratigraphy of the Burton Acres Shell Midden is the study of the layers in the site, the arrangement of those layers, and their content, into multiple sequences, the explanation of the origin of those sequences, and the correlation of those sequences with other sites in the region. The study begins during excavation by describing the deposits as they are uncovered and dividing them into meaningful subdivisions, called layers. The

layers are then interpreted in as much detail as possible and related back to the activities of the people that created them.

The site was excavated by dividing deposits into either *lithostratigraphic* or *arbitrary* (10cm) divisions. Lithostratigraphic divisions are based on the appearance of the sediment and the material making up the sediment. The appearance of sediment is called lithology, and refers to the physical appearance of the deposit. For the Burton Acres Shell Midden the lithology was primarily described in terms of the color of the fine-grained fraction (sand, silt and clay) and the composition of the large-grained fraction (shell, rock, bone, metal). Lithostratigraphic divisions were made whenever the deposits changed their appearance.

Divisions were also made on the basis of an arbitrary thickness. If a deposit based on lithostratigraphic criteria extended more than 10cm in depth, then the deposit was divided. This division was not based on lithology but rather on an arbitrary

---

Chapter opening photo: Elizabeth Martinson examines the layers exposed in the profile to record the stratification of the Burton Acres Shell Midden.

thickness, and thus is called an arbitrary division. For this site the thickness beyond which a layer should not extend was determined to be 10cm.

Before excavation began, three basic layers were identified by Stein from observations of the wave-cut bank. Portions of the subsurface were exposed in the nearby eroded bank, and could be examined as a preview of what would be found during excavation. The three layers were Layer 1 (sod), Layer 2 (cultural material), and Layer 3 (glacial till).

Each subgrouping of a layer (whether based on 10cm depth or physical evidence) was designated by an alpha-numeric name, such as 1A, 2A, 2B, 2C and so on. The sod layer was called 1A, and was composed of the root-mat and the grass plants. This deposit was removed as one piece of sod. It was a lithostratigraphic unit based on the presence of grass and roots, and it contained few artifacts. Layer 1 had only one sub-layer, "A". Layer 2 encompassed all the deposits that contained cultural material; most notably artifacts made of shell, charcoal, metal, and angular burned rock. Layer 2 deposits are referred to by archaeologists as shell midden, and are found in abundance on the shoreline of the Pacific Northwest. The Layer 3 was glacial drift, which is the sediment underlying all of Vashon and Maury Island that was laid down as the glaciers melted in this vicinity 15,000 years ago. Layer 3 deposits contained no artifactual material, and are referred to by archaeologists as "sterile."

EXCAVATION UNITS

Each excavation unit was 1m by 1m and named with the grid coordinates of the unit's southeast corner (see Figure 4.5 in previous chapter). Datum A was at 0 North/50 East. The units were designated: 22/58, 26/57, 28/58, and 29/58. The name reflects the location of each unit expressed in numbers of meters north and east of Datum A. Stakes were driven into the ground labeled with the grid coordinates.

Unit 22/58 was excavated differently from the other units. After excavating the first few levels, Stein determined that only the easternmost half of Unit 22/58 contained shell midden. The shell midden was so thin in this unit that no artifactual material was found in half of the unit. The decision was made to stop excavating in the western half of the unit and continue only in the eastern half. This changed the configuration of the unit

Figure 5.1 Unit 26/57 was excavated using methods different from those in the other excavation units. Only a few artifacts were found here, so the southeast quadrant (upper right corner) was the only portion of the unit excavated to depths greater than 5cm. This unit established the western boundary of the artifact distribution.

to 50cm by 100cm. This change in excavation technique was made at the beginning of Layer 2C, and was continued until the sterile layer, 3A, was reached. The greatest depth below surface was reached in the east half of the unit at 60cm below surface.

Unit 26/57 also was excavated differently. Almost no shell midden was found in this unit, as well as very few artifacts. Only one layer (2A) was excavated across this unit. Layer 2B was excavated from only the SE quadrant (25cm by 25cm section), and to a depth of 10cm below 2A (Figure 5.1). The unit was determined to be without artifacts. Layer 3A was sampled in only this same quadrant and in a volume of four buckets. The greatest depth below surface was reached in the southeast quadrant at 30cm below surface.

Unit 29/5E was opened at the same time as Units 22/58 and 26/57. Large quantities of shell midden were encountered in this unit, so an adjacent Unit 28/58 was opened. Each of these two 1m by 1m units was exca-vated until the glacial drift was encountered (Layer 3). In Unit 28/58, five layers were removed (1A, 2A, 2B, 2C, and 3A) to a depth ranging from 30 to 50 cm below surface. In Unit 29/58, eight layers were removed (1A, 2A, 2B, 2C, 2D, 2E, 2F and 3A) to a depth ranging from 50 cm to 65 cm below the surface.

LAYERS AND DEPTHS

The layers found in each unit are described below:

Unit 22/58 (Figures 5.2 and 5.3)

1A sod.

2A grey sediment with abundant roots, artifacts are present, but no shell. (36 liters).

2B tan sediment containing pebbly gravel. Looks like glacial drift that has been pushed up by a bulldozer. No artifacts are found. Layer contains large root of Madrona tree in the southeastern corner of the unit. (142 liters).

2C darker than 2B. Looks like it might be buried A soil horizon. Contains some shell (0-25%). Excavated in only the eastern half of the unit. (60 liters).

2D even darker sediment than 2C, and contains even more shell (25-50%) that is all fragmented. Obtained charcoal for potential radiometric dates. Observed no historic artifacts. (58 liters).

2E arbitrary layer division created after 10cm thickness was reached for 2D. Lithology similar to 2D. Madrona root in eastern half of unit. (48 liters).

2F similar shell midden as 2D and 2E, but with lower concentration of shell (25%). Close to bottom of midden. (22 liters).

2G tan sediment containing a few sparse shells and charcoal fragments. (16 liters).

3A glacial drift surrounding large roots of Madrona tree. (24 liters).

Unit 26/57 (Figures 5.4 and 5.5)

1A sod.

2A grey sediment with some roots, artifacts present, almost no shell. (102 liters).

2B changed to excavating only a 50cm by 50cm portion of unit in the southeastern quadrant. Tan sediment containing bits of shell and charcoal, one mammal bone, no artifacts and large root. End of shell midden. (36 liters).

3A glacial drift. (8 liters).

Unit 28/58 (Figures 5.6 to 5.10)

1A sod.

2A grey sediment with abundant shell (0-25%), bone, metal, and other artifacts, all fragmented and less than 1cm in size. Contains large amounts of rounded pebbles. This layer is equivalent to Layer 2A in adjacent Unit 29/58. (132 liters).

2B grey sediment with abundant shell (0-25%), bone, metal, and other artifacts. Contains rounded pebbles. This layer is equivalent to Layer 2B in adjacent Unit 29/58, but without the rounded pebbles found in Unit 29/5. (156 liters).

2C tan and grey sediment with abundant burned shell and corroded metal fragments. In some portions of the unit this layer represents shell midden that has filled pits excavated into the glacial drift. This layer is equivalent to Layer 2C in adjacent Unit 29/58. (112 liters).

3A glacial drift. (32 liters).

Unit 29/58 (Figures 5.11 to 5.13)

1A sod.

2A grey sediment with abundant shell (0 to 25%), bone, metal, and other artifacts, all fragmented and less than 1cm in size. Contains large amounts of rounded pebbles. This layer is equivalent to Layer 2A in adjacent Unit 28/58, but contains more rounded pebbles. (154 liters).

2B grey sediment with abundant shell, bone, metal, and other artifacts. Contains large amounts of rounded pebbles (25-50%) all between 1 to 3cm in size. This layer is equivalent to Layer 2B in adjacent Unit 28/58, but does not seem to represent filling of pits. (82 liters).

2C tan and grey sediment with abundant burned shell and corroded metal fragments. This layer is equivalent to Layer 2C in adjacent Unit 28/58, but does not seem to represent filling of pits. (164 liters).

2D grey and black sediment containing whole shell, and fire-cracked rock. Little metal. No equivalent in Unit 28/58. (96 liters).

2E grey and black sediment containing whole shell, and fire-cracked rock. Split from layer 2D because thickness greater than 10cm. Sediment is very moist with evidence of some organic preservation. No equivalent in Unit 28/58. (126 liters).

2F grey and black sediment containing whole shell, and fire-cracked rock. In some portions of the unit, this layer represents shell midden that has filled pits excavated into the glacial

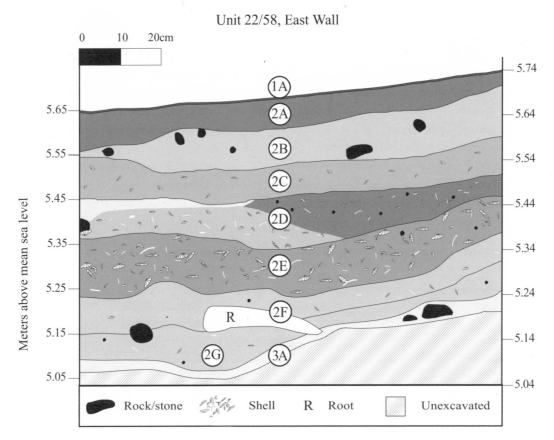

Unit 22/58, East Wall

Figure 5.2 Archaeologists usually represent strata as zones of similar looking sediments, drawing them as they appear in the walls of excavation units. In profiles, such as this one from Unit 22/58, each layer is shown in relation to all others. The ground surface is represented by the line at the top of the drawing, and the extent of the excavation is represented by the hatching at the bottom. The zones are then matched to the subgrouping of layers with the alphanumeric names of 2A, 2B, etc. In most cases the boundaries of the sediment zones correspond to the boundary of a named layer. But not always, such as in 2D/2E/2F.

Figure 5.3 Unit 22/58 was excavated differently from other units. Only half of the unit was excavated to a depth of 60cm, while the remaining half was left after only about 30cm of excavation. In this photo, the string divides the two halves of the unit with the far side representing the greatest depth of excavation.

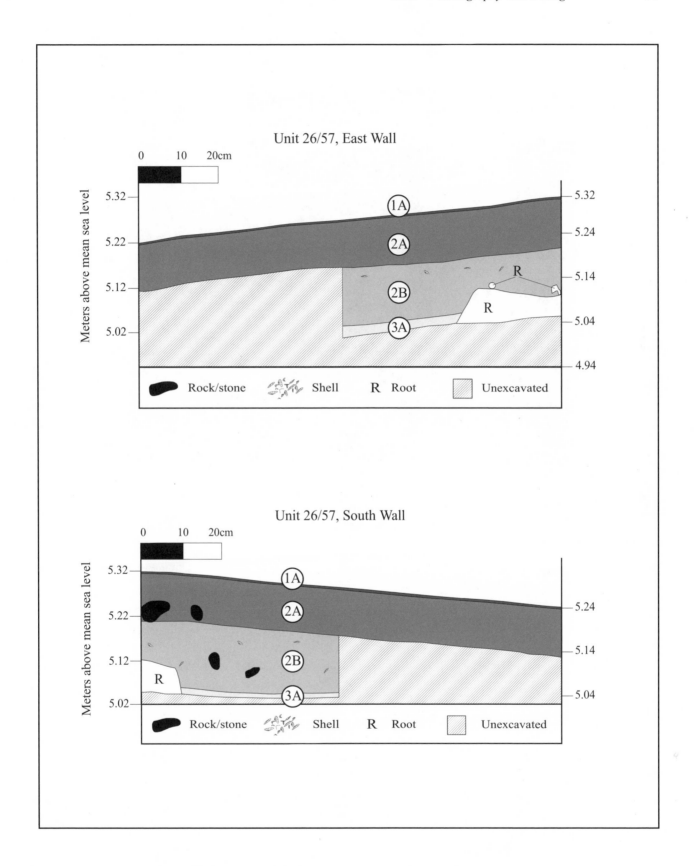

Figure 5.4 Unit 26/57 was only excavated to depth of 30cm, and even that depth was reached in only one quadrant. No shell midden was found in any layers of this unit, which established a boundary of the site. Since this location was beyond the boundary and contained almost no artifacts, excavation did not continue.

Figure 5.5 Unit 26/57 is shown here with only one quadrant (upper left, with string around it) excavated to 30cm depth. Note the location of Unit 22/58 in the background, a unit containing the shell midden. Evidence from these two units established the boundary of the site as between the two units. The archaeologists were surprised that the site did not extend further, suggesting that much of the site already must have eroded into the bay. The wave-cut bank is just visible in the upper left portion of this photo.

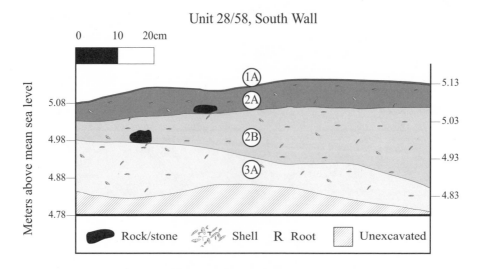

Figure 5.6 The layers visible in the south wall of Unit 28/58 were the thinnest of the whole unit. Just 10cm to 30cm of shell midden were found here.

Unit 28/58 and 29/58, East Wall

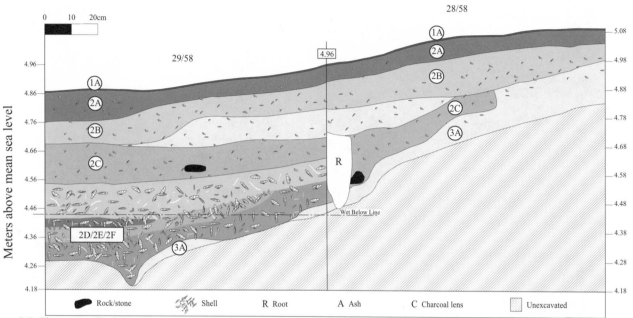

Figure 5.7 The east wall of the adjacent Units 28/58 and 29/58 shows the rapid increase in thickness of the shell midden from right to left. At the south wall of Unit 28/58 the shell midden is only 10cm thick and increases in thickness to 60cm at the north wall of Unit 29/58. Note the vertical feature labeled "R" (for root) near the center of the profile. Archaeologists are always careful to note anything that may disturb the context of artifacts. Roots often can be major sources of such disturbance.

Units 28/58 and 29/58, West Wall

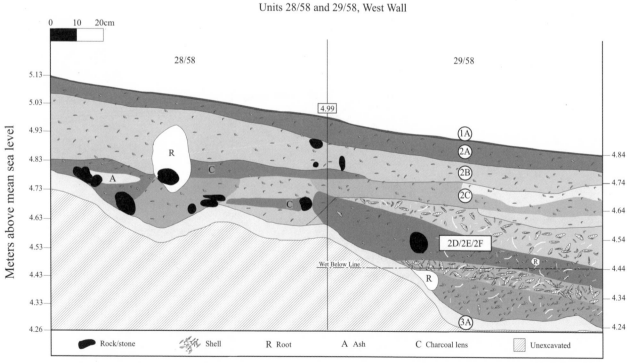

Figure 5.8 The west wall of Units 28/58 and 29/58 shows the relationship of the deeper and shallower deposits. The deeper layers (2D/2E/2F) contained older radiocarbon ages and almost no metal artifacts, and were interpreted as having been laid more than 500 years ago before contact with Europeans. The shallower layers (2A, 2B, and 2C) are interpreted to have been laid down after contact with Europeans because of the metal fragments and artifacts they contain. Note the configuration of the boundary of 3A along this profile. The steep drop-off is thought to have been carved by waves at an ancient shoreline later covered by people's occupational debris.

Figure 5.9 This view looking north shows the steep drop-off of the underlying glacial sediments. The drop-off was created by waves that eroded the bank more than 1000 years ago. People then lived on this shore and created more dry land as they deposited the by products of their occupation into the intertidal zone.

Figure 5.10 A close-up of the west wall of Units 28/58 and 29/58 shows the relationship between the underlying glacial deposits and the overlying shell midden. Fortunately, the excavation unit just happened to encounter the ancient buried shoreline at this spot.

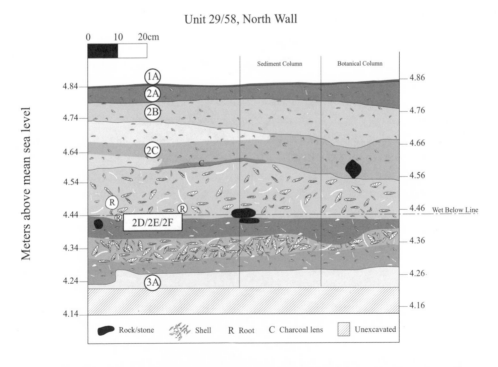

Figure 5.11 The north wall of Unit 29/58 was the location where two kinds of samples were taken, the sediment and botanical samples. Samples were extracted from the wall on the last day, excavating into the profile. The depth of each sample was recorded and later correlated with the layer name to which it belonged. Note that below elevation 4.45 meters above mean sea level the sediment was wet when excavated. The wetness was from brackish tidewater that infiltrated the base of the deepest part of the midden.

Figure 5.12 The matrix of the deepest layers is darker in color than the matrix of the upper layers. The color difference is caused by water infiltrating from below as the tide goes in and out. This water affects the preservation of the shell midden, dissolving shell and bone more rapidly than in the drier areas.

Figure 5.13 This view of the west wall of Units 28/58 and 29/58 shows the dark layer at the bottom and the light layer at the top. This dark layer at the base of the shell midden has been observed at other sites in the region. Also note the steep drop-off of the underlying glacial sediment just to the right of the arrow.

drift. Sediment is very moist with evidence of organic preservation (e.g., wood). No equivalent in Unit 28/58. (80 liters).

3A glacial drift.

CHRONOLOGY

The age of the layers uncovered at Burton Acres Shell Midden is determined by a combination of relative and radiometric dating. Relative ages are determined by noting the superpositional relationships of layers and considering the context of the artifacts found within them. Radiometric dating is the laboratory analysis of materials containing carbon to determine how much decay has occurred in those radioactive carbon isotopes. Both dating techniques are necessary because the age of an object determined by radiometric means is measuring the age of death of the organism, in which the carbon accumulated. That death may not correlate with the age of deposition for the layer. The context of the object within the layer is always evaluated carefully.

Relative Ages

The Burton Acres layers can be combined into two relative-aged groups: those that contain metal objects (and are therefore post-contact with Europeans), and those that do not contain metal and are, therefore, pre-

contact with Europeans (Table 5.1). These data suggest that the majority of layers excavated contained historic artifacts and were deposited by people living on the landscape after large amounts of metal were introduced through contact with Europeans. This does not suggest that Europeans were necessarily living in the vicinity or at the site, but that the people living in Quartermaster Harbor had acquired these goods through trade or direct purchase.

The age of these historic-period layers can be guessed from the age of manufacture of some diagnostic metal artifacts. A button, lead shot, and a dime suggest the period of occupation to be between AD 1860 and AD 1920 (see Chapter 6). Trade beads could suggest earlier occupation, and nails, glass, and plastic indicate later. These objects taken together suggest that most of these layers (other than the upper-most ones, e.g., 1A and 2A) were laid down late in the nineteenth century and early in the twentieth century.

Radiometric Ages

Fourteen samples were submitted to Beta Analytic, Inc. (Florida) for radiocarbon analysis (Figure 5.14; Table 5.2). Of these samples, seven were shell, six were charcoal, and one was uncharred wood. Samples were chosen from each unit (four from Unit 22/58; two from

Table 5.1 Pre- and Post-Contact Layers

Unit	Post-Contact	Pre-Contact
22/58	1A, 2A, 2B, 2C	2D, 2E, 2F, 2G, 3A
26/57	1A, 2A	2B, 3A
28/58	1A, 2A, 2B, 2C	3A
29/58	1A, 2A, 2B, 2C	2D, 2E, 2F, 3A

Unit 26/57; two from Unit 28/58; and six from Unit 29/58). In Table 5.2, both the measured carbon-14 age and the corrected age are provided. The carbon-14 age must be corrected to translate it into calendar years, which are usually expressed in years AD (or BC).

Another graphic display makes it is easier to compare the range of ages measured for each layer. In Figure 5.14, the ranges of radiocarbon ages (calibrated) are grouped by excavation unit. The samples within the units are arranged by depth and age. The depth is indicated on the X-axis (or right side of the three-dimensional column representing the unit). The date of the sample is arranged along the Y-axis (or bottom of the column representing the unit). The youngest samples are found on the left and the oldest toward the right.

Two samples suggest that the earliest occupation at Burton Acres was around 1000 years ago. A sample from Unit 22/58 indicates a radiocarbon age of between AD 690-890, or 1060-1260 years before present (BP). This charcoal date suggests that people deposited materials, notably a piece of burned wood, in this area of the shoreline over 1000 years ago. A charcoal sample from Unit 29/58 has a radiocarbon age of between AD 1035-1270, or 680 to 915 years ago (BP). This sample was collected from the most deeply buried layer, and together with the sample from Unit 22/58, points to an initial occupation (no matter how large) at just after 1000 years ago.

The rest of the samples indicate two periods of occupation based on the age-of-death for the tree or shellfish submitted for testing; one around 500 years ago and one within the last 300 years. Except for the two samples mentioned above, most of the shellfish and fish remains in the deeper layers of Units 29/58 and 28/58 were deposited 500 years ago. This conclusion stems from the fact that multiple radiometric samples returned results of that time period, and that the samples were

extracted from a wide variety of depths. Samples in Units 26/58 also clustered around 500 to 600 years ago.

The most recent occupation, from the present back to about 300 years ago, is more accurately dated from artifactual material than radiocarbon dating. The age-of-manufacture for the button and seated-liberty dime provide windows of time that are shorter than is the range of dates provided from radiometric dating. Few samples from the layers containing metal were submitted for analysis, because the age of those deposits was known with some certainty. Some dated charcoal did return ages within this historic period (sample from Units 28/58 and 29/58), thus corroborating the evidence provided by the artifacts.

These radiometric dates support the artifactual evidence that people lived at Burton Acres during the time before and after contact with Europeans, and that Native American were long-time, continuous, residents of the harbor. The exact residence style of the occupants of this site is not known. People may have come seasonally to this point to collect shellfish and fish. Today the wind affords excellent opportunities for drying anything. Alternatively, people may have come to the point and stayed permanently over all seasons, moving on to other locations only when resources were depleted seasonally or when other locations offered greater advantages. The contents of the deposits uncovered in this excavation suggest that the occupants occupied the location, and processed resources that became abundant seasonally. More deposits, however, would have to be uncovered to answer the question with certainty.

INTERPRETATIONS OF LAYERS

Once the dirt in the layers has been described and dated, the interpretation of individual layers' sources and deposition can be made. Archaeologists usually do this chronologically from oldest to youngest, describing reconstructions of the landscape and environment based on the clues in the layers they found. The reconstructions of the Burton Acres Shell Midden site are arranged for convenience into headings of the *glacial period*, *1000 years ago*, for which little information is available, *500 years ago*, when most of the shell midden in the deepest unit was deposited, and *200 years ago*, when large amounts of material were spread over the area. Also included is a *modern period*, when more current

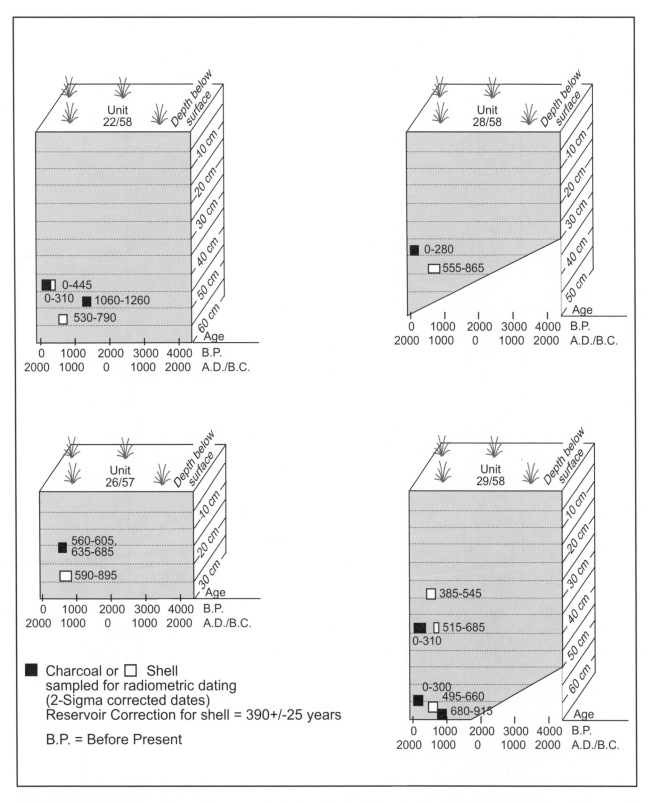

Figure 5.14 These blocks represent the excavation units with the depth of each radiometric sample shown relative to the surface. The position of samples (right to left) is relative to the time scale at the base of the block. The thickness of the sample box represents the precision of the radiometric dating technique expressed in years. The numbers next to each sample box is the age range (B.P. or before present) obtained by radiometric dating of each piece of charcoal or shell (see Table 5.2 for details). This graphic arrangement demonstrates that much of the lower half of the shell midden accumulated rapidly between 1000 and 500 years ago, and the upper half accumulated after about 300 years to the present.

Table 5.2 Radiocarbon Dates for Burton Acres Shell Midden

UW Sample Number	Beta Analytic Number	Material*	Measured C-14 Age (BP)	C13/C12 Ratio (o/oo)	Conventional C-14 Age (BP)	Reservoir Correction	Intercepts of Radiocarbon Age with Calibration Curve (AD)	1-Sigma Calibrated Results (AD)	2-Sigma Calibrated Results (AD)	Depth (cm)	Unit
22 2F 51	96004	Shell	380+/-190	0.0	790+/-200	390+/-25	None	1695-1950	1505-1950	43	2258/2F/005
22 2F 54	96005	Charcoal	140+/-80	-25.0	140+/-80	N/A	1690, 1735, 1815, 1925	1665-1950	1640-1950	43	2258/2F/005
22 2G 244	101897	Charcoal/AMS	1200+/-40	-23.3	1230+/-40	N/A	790	770-875	690-890	49	2258/2G/002
22 3A 31	96006	Shell/EC	700+/-80	0.0	1110+/-80	390+/-25	1295	1240-1345	1160-1420	53	2258/3A/003
26 2B 181	96007	Shell	760+/-80	0.0	1180+/-80	390+/-25	1250	1170-1300	1055-1360	23	2657/2B/018
26 2B 442	101898	Charcoal/AMS	680+/-40	-23.6	700+/-40	N/A	1290	1280-1300	1265-1315, 1345-1390	16	2657/2B/004
28 2C 942	101899	Charcoal/AMS	70+/-50	-22.5	110+/-50	N/A	1825, 1835, 1880, 1915	1680-1745, 1805-1935	1670-1950	34	2858/2C/009
28 3A 11	96008	Shell	730+/-70	0.0	1150+/-80	390+/-25	1275	1200-1315	1085-1395	39	2858/3A/001
29 2C 671	96009	Shell	450+/-60	0.0	860+/-60	390+/-25	1465	1435-1505	1405-1565	28	2958/2C/067
29 2D 51	96010	Shell	620+/-60	0.0	1040+/-60	390+/-25	1335	1300-1405	1265-1435	38	2958/2D/005
29 2D 544	101900	Charcoal/AMS	250+/-60	-29.2	180+/-60	N/A	1675, 1770, 1800, 1940	1660-1700, 1720-1820, 1855-1860, 1920-1950	1640-1950	38	2958/2D/005
29 2F 80	96011	Wood/EC	120+/-80	-25.0	120+/-80	N/A	1700, 1720, 1820, 1855, 1860, 1920	1670-1780, 1795-1945	1650-1950	60	2958/2F/008
29 2F 81	96012	Shell	590+/-60	0.0	1000+/-60	390+/-25	1390	1320-1425	1290-1455	62	2958/2F/008
29 2F 84	96013	Charcoal	870+/-50	-25.0	870+/-50	N/A	1195	1065-1075, 1155-1235	1035-1270	65	2958/2F/008

* EC = sample required extended counting; AMS = sample required accelerator-mass-spectrometer analysis.

evidence can be obtained from informants, and aerial photos of bulldozing and house construction.

Glacial Period

During the Ice Age, and until about 15,000 years ago, a thick mass of ice covered Puget Sound (Thorson 1980; Waitt and Thorson 1983; Booth 1984). As this ice melted, the region was blanketed with sediments carried within the ice from places further north. The sediment is referred to as glacial till or drift, and is characterized by a wide variety of particle sizes, ranging from clay to boulders. Vashon Island is covered with this glacial drift ranging from a massive layer of fine-grained sediment such as is exposed in the bluff just south of Burton Acres Park, to more well-sorted gravels found on the eastern shores of Vashon Island.

The glacial sediments settled as the ice melted and created undulating surfaces on which soils developed. In these soils a succession of plants colonized. Initially, a mixture of tundra and parkland vegetation covered the area with spruce, alder, pine, mountain hemlock, with Western hemlock dominating (Whitlock 1992:10; Hebda and Whitlock 1997). From 9000 to 5000 years ago the climate was warmer and drier, resulting in a more open forest or savannah that included Douglas firs and oaks. After 5,000 years ago the climate returned to a wetter environment, and Western red cedar spread into the region. The forests we envision for the period before modern logging, therefore, did not arrive in the region until probably around 5000 years ago (Hebda and Mathewes 1984). Those trees survived the Ice Age in the non-glaciated portions to the south, and spread slowly northward one seed and one tree at a time. Thousands of years passed until the appropriate soils developed and seed sources arrived.

The glacial deposits underlying the Burton Acres Shell Midden site represent the melting of the ice and the developing of soil on the stable landscape thus created, and contain no evidence of people. The glacial drift and the zone of soil formation (Layer 3A) represents many thousands of years of stability, when no people dropped material on this particular surface. People were walking over other surfaces of Vashon and Maury Islands for all these thousands of years because residents of the islands brought artifacts for the archaeologists to identify that were fashioned in a style made as long ago as 5000 to 9000 years. One such object was found on Maury Island just across

Quartermaster Harbor from Burton Acres Park. Thus, even though we found no evidence of people living at Burton Acres Shell Midden between 11,000 and 1000 years ago, they must have been nearby.

1000 years ago

Quartermaster Harbor and Vashon Island are very different now from their appearance 1000 years ago. The vegetation and fauna, shoreline configuration, and tidal flow were all different in subtle and complex ways. The land was covered with old growth forests in the interior, with understories of shrubs and smaller trees. Native people describe berries as important resources on the land as well as large numbers of deer populating the forests.

Quartermaster harbor was vastly different, because Maury Island was separated from Vashon Island at the location of Portage, allowing water (and fish) to flow at high tide from the open straight into the harbor. The harbor was 'flushed' with every high tide bringing volumes of water into the north end of the bay and forcing it out the south end. The low isthmus at Portage existed for at least the last 1000 years, until a road was built in this century. Many informants observed salmon waiting for high tide to cross the isthmus and swim into Quartermaster Harbor to spawn in Judd Creek.

Herring were another important inhabitant of Quartermaster Harbor. Vast numbers of herring swam north every spring to spawn in the shallows of the harbor's shorelines. Residents of the island, both Euro-American and Native American, remember swarms of herring running north so close to the point at Burton Acres Park that you could throw a net from the shore and catch them (Figure 5.15). These runs continued until the habitat was destroyed in the last few decades.

Other fish and shellfish species remembered by residents are no longer found in Quartermaster Harbor because of pollution and habitat destruction. Not only the creation of the dam at Portage but the building of seawalls, dumping of pollutants, and blocking of creeks have altered the diversity of life in the harbor today.

Besides human-induced changes to the area, natural catastrophes have affected Quartermaster Harbor. In Puget Sound, a profound earthquake occurred 1100 years ago causing massive earth movement along the Seattle Fault (Atwater and Moore 1992; Bucknam et al. 1992; Jacoby et al. 1992; Johnson et al. 1994; Karlin and Abella 1992; Schuster et al. 1992). Stretching from

Figure 5.15 This aerial photo taken in 1944 shows the Burton Acres Park as a white point of land projecting east (right). The herring migrated to the north and west, swimming close to this point on their way around the Burton peninsula to the northern shores of Quartermaster Harbor.

Bainbridge Island through southern downtown Seattle and east to Issaquah, the Seattle Fault was the dividing line for land movement during this great earthquake (defined as greater than Richter-scale 9). North of the fault, the land was suddenly down-dropped by about three meters (10 feet) for a distance of a few miles. South of the fault the land was raised by 3 meters to 6 meters (10 feet to 20 feet). This movement generated a tsunami within Puget Sound that traveled northward ripping up sands from the intertidal zones and spreading sediment high onto beaches as far north as Whidbey Island.

Brian Sherrod (1998) believes Vashon Island to be beyond the southern limits of the uplift created by the seismic event along the Seattle Fault because he found no evidence in a pollen core taken from a wetland near Portage on Vashon Island. The plants growing in the wetland from which the core was taken show no evidence of being emerged from a brackish environment to a dry raised setting. He cautions that small-scale movement may be difficult to detect in the core, if the shifting of the land was slight, and insufficient to lift the surface beyond the effects of the saltwater. People living in Quatermaster Harbor would have felt this seismic event, although its long-term impact may have been minimal.

The Burton Acres Shell Midden site 1000 years ago extended further into the harbor than it does today. Erosion has removed large portions of the bank in the last 10 years, and residents of the island informed the archaeologists that, years ago, the bank extended beyond its present position. The configuration of this eroded landscape is unknown, but one can speculate that it formed an extension of the high bluff that rises to the south.

A fresh-water stream entered the harbor at Burton Acres Shell Midden. A close-up of the 1936 aerial photo of the area shows a wetland nestled up against the bluff enclosed by a berm at the shoreline (Figure 5.16), and a stream entering the wetland at the location of what is now the parking lot for the boat ramp. Salmon may have migrated up the stream to spawn and the banks may have provided habitat for fresh-water mussels (a few specimens of which were found in the site, see Chapter 11). The wetland was probably filled with brackish water for most of the year and may have supported shellfish. The wetland was confined by a berm on the east side that seems to have remained above sea level even at high tide. The relationship of the berm/wetland/stream may have shifted in the past as sediment was eroded from the high bank to the south, as the land moved in relation to earthquakes, and as rare extreme storm events shifted the berm sediments.

The reconstruction for this period suggests that people occupied the point of land created by the berm and the northern extreme of the bluff. They either stayed there for only a short time (leaving only a few artifacts) or they occupied an area now eroded and destroyed. The excavation units in our dig were placed in a location that encountered only a small amount of material from this period. Other areas of Puget Sound were densely occupied at this time, so we know there were large populations of people on the island as well as in other parts of Puget Sound. Our interpretations are that this area, too, was occupied intensively, but that the deposits either have been destroyed or simply have not been encountered in this excavation.

The occupation that occurred 1000 years ago is known from only two radiocarbon dates found very close to dates of 500 years ago. The exact amount of remains in these deposits laid down in the years between 1000 and 500 years ago is not known. The important point is that (other than the radiocarbon ages) the deposits themselves look very much alike. They probably represent similar activities over this period of time.

500 years ago

As mentioned above, between 1000 and 500 years ago no changes are detected within the layers of the site. The deposits at Burton Acres Shell Midden contain charcoal samples of different ages, but similar shellfish, bones from salmon, herring, deer, and other animals. These pre-contact deposits (deposits that were laid down before contact with Euro-American people) are found in three of the four units excavated (Table 5.1). They are characterized by higher numbers of whole shellfish (as opposed to fragmented and burned) and the absence of historic (especially metal) artifacts.

In the northern-most excavation unit (Unit 29/58) these pre-contact layers were differentiated by more than just larger percentages of shell. The sediment surrounding the shell was darker in color, almost black, and much wetter than all other deposits at the site. The dark color is caused by moisture controlled by the tidal fluctuation, a fact documented at the English Camp (British Camp) site on San Juan Island (Stein 1992, 1996, 2000). The position of the layers and the fact that they were periodically wetted allows us to interpret the changes in the landscape that have occurred between then and now.

The berm on which people were living 500 years ago was a slightly different shape from what it is now. People were living on a surface that was south and east of the area covered by grass today (Figure 5.17). Sea level 800 years ago was approximately at the same level as it is today. Because we find artifacts below the level of high tide today, we can assume they were throwing shellfish, bones, and charred plant remains down the front of the berm (toward the water) onto a slope that was covered by water in high tide.

Another explanation for the wetting of these lowest layers in the site is that sea level rose, rather than that the remains were being thrown into the intertidal zone. This explanation is problematic, in that geologists state that south of the Seattle fault the land was raised. Vashon Island is likely beyond the effects of this raising, but a small rise is far more likely than a drop of land relative to sea. Both explanations could result in the wet deposits we found, but the tossing onto the front of the berm is more likely given the geological events that occurred.

Figure 5.16 This close-up of a 1936 aerial photo of Vashon Island shows that Burton Acres Park was then forested. A wetland drained into the harbor. Fallen tree trunks are visible as crisscrossing lines at the edge of the wetland. The point has been turned white by shells eroding from the midden.

The wetting of the deposits resulted in excellent bone and charcoal preservation, as well as wood. In fact, two pieces of uncharred wood were found, and were interpreted originally to be the spectacular results of inundation and unusual preservation. The radiocarbon date of one of these uncharred wood samples was, however, determined to be from a recent period between AD 1650 and AD 1950, in other words post-contact. The wood must have been driven down into the underlying unit at some point in the past 300 to 50 years. It most likely was a stake associated with herring fishing and setting of nets. The wood was not contemporaneous with wetted layers and, therefore, was not evidence of the occupation.

Even with the disappointment of the wood preservation, the shellfish, mammal, and fish bone were spectacularly preserved. Tiny elements of fish skulls and interior periostraca of the most fragile mussels were all preserved. This helped greatly with identification and delighted everyone who assisted in the sorting of these amazing finds.

200 years ago

The deposits identified as being laid down sometime within the last 200 years were characterized by abundant amounts of metal and evidence of intense

Figure 5.17 Kate Donahue is standing in front of the eroded bank at the Burton Acres Shell Midden. About 500 years ago people were living on the surface behind her head. This surface extended seaward for an unknown distance. The Madrona tree to Kate's right that is about to fall onto the shore and further erode the bank.

Figure 5.18 This aerial photo taken in 1960 shows the glacial drift (white) exposed at the shore and the road curving upslope (down photo) to the west away from the point of the beach. Construction of this road resulted in glacial drift being pushed over the midden, preserving the shell and artifacts.

burning. The thickest deposits were in the northern two excavation units (Units 28/58 and 29/58) at the edge of the berm closest to the water. The people living at the Burton Acres Shell Midden at this time were using the berm as a platform for large hot fires. The bone and shellfish found in these layers was heavily fragmented and burned. Some of it was so burned, that it fell through the smallest screen as flecks of ash.

The metal found was extremely weathered making identification difficult. Much of it was rod-shaped and found in a checkered pattern laying parallel and perpendicular to each other. The pieces suggested the remains of a grating that had been laid on the fire and burned beyond recognition. The remnants of metal were so fragmented, however, that such an interpretation was perhaps wishful thinking rather than reality.

No feature resembling a defined hearth or fire-pit was found. The layers filled with ash, charcoal, burned shell and bone, and weathered metal spread across the unit in varying thickness and extent. If the particles were once part of any discreet hearth, they must have been removed and scattered over the surface. Interesting to us was the inclusions of so many fish bones and shellfish. The occupants of the site must have tossed

large numbers of fish (mostly herring) and shellfish into the fire while it was burning hot.

Members of the Puyallup Tribe of Indians remember their relative Lucy Gurand living at the location that is now Burton Acres Park. She lived in a 'house boat' tied to the shore. Perhaps it was she and her relatives that built the fires on the berm, and threw the results of their harvest into the fire, spreading later to form a new surface on which to process more herring and shellfish. The button from a uniform and a dime could have been theirs (see Chapter 6).

Modern

Although residents of the island remember events in the recent past, few people describe these events in sufficient detail to explain the layers that archaeologists discover at sites. At Burton Acres in the two southern units (Units 22/58 and 26/57), located highest on the slope and closest to the eroded disturbed bank, we found layers above the shell midden that seemed to have been moved within the last few years. The most conspicuous layer was glacial drift, looking very much like the Layer 3A found at the base of the unit (and the bottom of the exposed wave-cut bank). This layer resting near the present surface and above the shell midden had no

Figure 5.19 The bank was eroding so close to Unit 22/58 that shoring was built to protect it during the excavation. We feared that people walking near the edge would cause the bank to collapse. After the excavation units were back-filled, the shoring was removed. Residents inform us that a small amount of collapse has occurred since 1996.

shellfish, bone, or charcoal, was tan in color and contained a variety of particle sizes from clay to cobbles. Yet, it was on top of the layers containing artifacts.

The aerial photo from 1960 (Figure 5.18) provided us with the explanation for this deposit's origin. While homes were being constructed on the top of the adjacent bluff a road had been constructed to assist with the access to the building site. The road was built by a bulldozer pushing glacial drift from the west side of the bluff, toward the east and covering the site layers containing artifacts. The road can be seen in the aerial photo and when the people living in nearby houses were questioned, they all remembered the road and the bulldozer.

Inadvertent burial of archaeological sites is sometimes beneficial for preservation. In this case the bulldozer did not scrape layers containing artifacts and destroy their context. We were fortunate that by accident the bulldozer operator pushed glacial drift onto the site and saved a portion of it.

Lastly, for some reason in the past few decades, the bank at this location has begun to erode (Figure 5.19). Perhaps it has been eroding for some time and we only have recollections that go back a few decades, but this

erosion has destroyed an unknown portion of the site and continues to threaten the remainder.

REFERENCES

Atwater, B.F., and A.L. Moore
1992   A Tsunami about 1000 Years Ago in Puget Sound, Washington. *Science* 258:1614-1617.

Booth, D.B.
1984   Glacier Dynamics and the Development of Glacial Landforms in the Eastern Puget Lowland, Washington. Unpublished Ph.D. dissertation, Department of Geology, University of Washington, Seattle.

Bucknam, R.C., E. Hemphill-Haley, and E.B. Leopold
1992   Abrupt Uplift Within the Past 1700 Years at Southern Puget Sound, Washington. *Science* 258:1611-1614.

Hebda, R.J., and R.W. Mathewes
1984   Holocene History of Cedar and Native Indian Cultures of the North American Pacific Coast. *Science* 225:711-713.

Hebda, R.J., and C. Whitlock
1997   Environmental History. In *The Rainforests of Home, Profile of a North American Bioregion*, edited by P.K. Schoonmaker, B. von Hagen, and E.C. Wolf, pp. 227-254. Island Press, Washington, D.C.

Jacoby, G.C., P.L. Williams, and B.M. Buckley
1992   Tree Ring Correlation Between Prehistoric Landslides and Abrupt Tectonic Events in Seattle, Washington. *Science* 258:1621-1623.

Johnson, S. Y., C.J. Potter, and J.M. Armentrout
1994   Origins and Evolution of the Seattle Fault and Seattle Basin, Washington. *Geology* 22: 71-74.

Karlin, R.E., and S.E.B. Abella
1992   Paleoearthquakes in the Puget Sound Region Recorded in Sediments from Lake Washington, USA. *Science* 258:1617-1620.

Salvador, A.
1994   *An International Stratigraphic Guide: A Guide to Stratigraphic Classification, Terminology, and Procedure*. 2nd ed. Geological Society of America, Boulder, Colorado.

Schuster, R.L., R.L. Logan, and P.T. Pringle
1992   Prehistoric Rock Avalanches in the Olympic Mountains, Washington. *Science* 258:1620-1621.

Sherrod, B.L.
1998   Late Holocene Environments and Earthquakes

in Southern Puget Sound. Unpublished Ph.D. dissertation, Department of Geology, University of Washington, Seattle.

Stein, J.K.

1987 Deposits for Archaeologists. In *Advances in Archaeological Method and Theory* vol. 10, edited by M.B. Schiffer, pp. 337-393. Academic Press, Orlando, Florida.

1990 Archaeological Stratigraphy. In *Archaeological Geology of North America*, edited by N.P. Lasca and J. Donahue, pp. 513-523. Geological Society of America, Centennial Special Vol. 4. Boulder, Colorado.

1992 Sediment Analysis of the British Camp Shell Midden. In *Deciphering a Shell Midden*, edited by J.K. Stein, pp. 135-162. Academic Press, San Diego.

1996 Geoarchaeology and Archaeostratigraphy: View from a Northwest Coast Shell Midden. In *Case Studies in Environmental Archaeology*, edited by E.J. Reitz, L.A. Newson, and S.J. Scudder, pp. 35-54. Plenum Press, New York.

2000 *Exploring Coast Salish Prehistory: The Archaeology of San Juan Island*. University of Washington Press, Seattle.

Thorson, R.M.

1980 Ice-Sheet Glaciation of the Puget Lowland, Washington, During the Vashon Stade (Late Pleistocene). *Quaternary Research* 13:301-321.

Waitt, R.B., Jr., and Thorson, R.M.

1983 The Cordilleran Ice Sheet in Washington, Idaho and Montana. In *Late Quaternary Environments of the United States*, vol. 1, edited by H. E. Wright, Jr., pp. 53-70. University of Minnesota Press, Minneapolis.

Whitlock, C.

1992 Vegetational and Climatic History of the Pacific Northwest During the Last 20,000 Years: Implications for Understanding Present-day Biodiversity. *The Northwest Environmental Journal* 8:5-28.

6

Historic Artifacts

MaryAnn Emery

Some objects found in archaeological sites were made so recently that written descriptions of their manufacture and use are available. From these documents we know their age and function. Artifacts manufactured recently, within the historic past, are described in this chapter. Although we know many things about these objects, we still are not sure exactly who used them or why they were discarded at this site.

Archaeologists divide the discipline of archaeology into two main categories, which they identify as prehistoric and historic. Prehistoric archaeology studies the period of time during human history where there are no written records, and historic archaeology studies the period for which there are written records to supplement the archaeological record. This boundary between prehistory and history is not always clear, because it occurs at different times in different parts of the world, depending on the development of writing systems or the introduction of writing from outside (such as the Romans in Britain). As a result, in some cases the records kept are from the perspective of an outside culture and not the culture that is being studied archaeologically. A further complication arises because many aspects of the culture are not recorded in the early stages of a writing system or when an outside source is doing the recording. For instance, in many cultures economic transactions were the only records kept, which does not tell us very much about the culture as a whole.

This fluctuating boundary and limited recording procedures pose special problems for North American archaeology.

In North America, historic archaeology is associated with the time when any form of contact occurred between Native Americans and Europeans. On the east coast this was generally a face-to-face contact immediately or soon after an incidental contact (through trade for European goods or the spread of disease). In the Northwest, contact occurred first through the introduction of European manufactured goods that were traded into the region from other Native Americans who had face to face contact with Europeans several hundred years before the peoples of the Northwest coast had direct contact with them. Therefore there is a long period of time in the Northwest after "contact" but before written records. As a result, the initial contact period is not well documented and is also often ignored by prehistoric archaeologists as being "historic." Thus, this time period has not been extensively studied and is

Chapter opening photo: Laura Andrew, a University of Washington archaeologist, catalogs an historic artifact recovered by a volunteer in his 2-liter bucket.

not well understood. Many questions remain as to the exact nature of the contact between Euro-Americans and Native Americans on the Northwest coast, especially the effects before "face-to-face" contact on daily lives of the people of the Northwest coast.

The excavation at the Burton Acres Shell Midden site on Vashon Island may help to answer some of those questions. The presence of European or American (Euro-American) artifacts in a layer of the Burton Acres Shell Midden was used as a marker to identify that layer as post-contact. This layer could then be compared to the pre-contact layers to look for changes or stability in subsistence practices and technology. The analysis of these historic artifacts was undertaken with two main goals in mind. The first goal is to determine the date of manufacture of the artifacts to obtain an approximate date when contact occurred at Burton Acres Shell Midden. The second goal is to define the probable function of the artifacts to reconstruct the activities at the site. Finally, after each artifact is analyzed and the data is recorded, quantitative analyses are undertaken to determine if any patterns exist in the archaeological record.

Another aspect of the historic component at the Burton Acres Shell Midden is related to the manner in which it was excavated. Many historic sites are excavated in large arbitrary layers because it is believed that historic sites are deposited so quickly that no information can be obtained from excavating in natural or smaller layers. The precise provenience control and small layers used at Burton Acres Shell Midden will provide data that can test this hypothesis.

HISTORY OF CONTACT

Before examining the historic artifacts in detail, a brief review of the recorded accounts of contact will provide a foundation for examining the historic artifacts. According to published historical records, the first known face to face contact between the people of the southern Puget Sound and Europeans was in 1792 when George Vancouver entered the Puget Sound (Suttles and Lane 1990). Suttles and Lane report that Vancouver found evidence of European disease and artifacts, but that it did not appear that the Native Americans had encountered Europeans before. Very little known contact occurred between the time of Vancouver's voyage, and the establishment of Hudson's Bay Forts during the

1820s and 1830s that serviced the fur trade in the Columbia District (Suttles and Lane 1990). Fort Nisqually (the closest fort to Vashon Island) was established in 1833 and existed as a trading center with store until 1860 (Steele 1977). By 1853 several Euro-American communities in the Puget Sound were prospering and the Washington Territory was established. Cole and Darling (1990) suggest that the lives of Native Americans did not change drastically until the establishment of the European community.

The Hudson's Bay Company provided European goods but they did not interfere to any great extent with the day-to-day lives of the Native Americans except to pay them for furs and to employ a small number at the forts and as voyagers. Chance (1973) reports that for the Native Americans living near Fort Colville on the Columbia River Plateau in the early years of the Hudson's Bay Company the main change in their lives was the introduction of metal as a raw material. Chance suggests that the Native Americans still hunted and fished in the traditional manner and in the traditional areas, however, they used metal fishing hooks, projectile points, and knives. Although there were attempts by the Hudson's Bay Company to influence the lives of the Native Americans in other areas, it was only with the arrival of missionaries, settlers and the reservation system that the lives of the Native Americans changed drastically.

Analysis of the historic artifacts found at the Burton Acres Shell Midden gives us clues as to the nature of the contact between Native Americans in the southern Puget Sound, when this contact may have occurred, and the nature of the changes that may have taken place after contact.

METHODOLOGY

Sampling

One hundred percent of the historic artifacts from all units are analyzed.

Sorting

Historic artifacts with the same provenience (recorded at the scale of 2- or 8-liter buckets and one of four screen fractions) are sorted by material class; each material class is in a separate bag with an identification label on the outside of the bag. The material classes for historic artifacts are: glass, metal, ceramic and a general historic category that contained leather, textiles, rubber, plastic,

Table 6.1 Markings and Manufacture Dates for .22 Casings

Stamp	Manufacturer	Dates of Operation
US	United States Cartridge Co.	1868-1935
U	Union Metallic Cartridge Co.	1867-1910
U	Remington Arms-Union Metallic Cartridge Co.	1910-present
PETERS HV	Peters Cartridge Co.	1887-present

After Plew et al. (1984).

fiberglass and Styrofoam. Each artifact is identified to artifact type (i.e., bottle, bullet casing). In addition, specific attributes of each artifact and notes about the condition of the artifact are recorded (Appendix C).

RESULTS: DESCRIPTIVE

This section is a descriptive analysis of the identifiable artifacts, including date of manufacture if known, and artifact function if known.

Metal

Artifacts relating to hunting/guns make up the largest category of identifiable metal artifacts. There were 10 bullet casings or cartridges found. Nine .22 caliber casings and one .30-30 Remington casing. The markings on the bases of the .22 casings include: "SUPER X", "HP", "PETERS HV", "U" and "US". Table 6.1 lists the manufacturers and dates of most of these cartridges. All of the .22 casings were found in Layer 2A of Units 29/58 and 28/58.

Seven percussion caps and three pieces of lead shot were found in Layers 2B and 2C of Units 29/58 and 28/58. The percussion cap was invented in 1814, and was used with lead shot in a gun commonly called a musket (Johnson 1943) (Figure 6.1). A pin-fire cartridge case and lid was found in Layer 2C of Unit 29/58 (Figure 6.2). The pin-fire device was one of the first self-exploding cartridges and was quite revolutionary in ammunition. It was invented by Le Faucheux of Paris in 1836 and was patented by Houllier in 1846 (Johnson 1943; Serven 1964). This cartridge combined the gunpowder, priming element, projectile, and extraction device in one. One of the unique features was a mechanism for sealing the bore against the escape of powder gases to the rear, which was a big improvement for the health of the shooter. However, the fragility of the cartridge and the tendency to fire accidentally

Figure 6.1 This example of a percussion cap indicates that a musket was fired at the site. Seven of these caps, which were invented in 1814, were found in the upper-most layers of Units 28/58 and 29/58. These caps, along with abundant shattered glass, suggest that someone may have shot at glass bottles for target practice. Cap is 13mm at widest point.

Figure 6.2 This rarely-found pin-fire cartridge case and lid was one the first self-exploding cartridges, invented by LeFaucheax in Paris in 1836. Because the cartridge occasionally fired accidentally its use was restricted to just a few decades. Excavations at Kanaka Village, located outside of the Hudson Bay Company's Fort Vancouver, recovered a similar cartridge.

limited the life span of the pin-fire cartridge to a few decades. A similar cartridge was found at Kanaka Village, the housing area for the Hudson's Bay Company workers at Fort Vancouver. The cartridge from Kanaka village was excavated from the Quartermaster's

Figure 6.3 Many nails were found in the excavations, almost all of those identified were brass, machine-headed nails. When nails are found in archaeological sites they are usually interpreted to have been part of a building or furniture, rather than as loose nails simply dropped on the ground. These nails were all concentrated in a burned area, which may signify that they were part of furniture or a chest that was broken up and burned as fuel in a fire.

Figure 6.4 This dime is so corroded that it could only be identified by shining light obliquely across its surface while looking under a microscope. From the light an image of a seated Liberty figure emerged. The minting date for this figure is between 1860 and 1891. The date on this specimen had been worn away completely.

house, and was found with materials dating from 1850 to 1900 (Chance and Chance 1976).

Another possible line of evidence that may indicate the use of guns at the site requires a brief look at the glass assemblage. The glass at the Burton Acres Shell Midden site is highly fragmented. This fragmentation of the glass and the presence of flake scars, possibly from bullet impacts, on several pieces of glass, suggests that some of the glass may have been broken during target practice. There is even one piece from Layer 2A of Unit 29/58 that has flakes removed from two sides which could have resulted from being shot from both sides.

The other identifiable metal artifacts found at the site include several burned, "early machine headed cut nails" (Figure 6.3). These nails are manufactured of brass which does not corrode as quickly as iron, and thus were more easily identifiable. Nelson (1968) describes these nails as produced from the 1810s to the late 1830s, having the following characteristics:

"Nails of this period are distinguished by
their irregular heads, which vary in size
and shape, usually eccentric to shank,
though they were more uniform by 1830's.
Nails were irregular in length and width,
but more uniform at end of period. Nails

generally have a rather distinct rounded
shank (under head), caused by wide
heading clamp."

These nails were often used as finishing nails and may have come from a piece of furniture. Since furniture and nails in general have a long artifact lag time (the time between manufacture, and the average time it takes for the artifact to be deposited in the archaeological record) it is difficult to determine when the nails were used and deposited at this site. Most of these nails were found in Layer 2B of Unit 28/58. Since all of the nails were found in the same general area of the site it seems more likely that they came from a piece of furniture. There was also evidence of burning in this layer and the nails could have come from a chest or other piece of furniture that was burned. The rest of the nails were too corroded to identify to type or to date the time of manufacture. There were a few nails that could be identified only as square or round.

Three coins were also among the metal artifacts excavated. Two pennies dated 1967 and 1970 and one dime were found. The dime (Figure 6.4) was almost unidentifiable. Most of the markings were worn away.

Figure 6.5 This thin metal hinge may have been used to decorate an important book or a drawer on a piece of furniture. Nothing else like it has been found at other historic archaeological sites in the region. It could have decorated any number of objects, but its shape resembles a hinge.

Figure 6.6 A Prosser-type button (left) was found in Layer 2C of Unit 29/58. Two seed beads, one cobalt blue and the other white, were found in Unit 29/58 in Layers 2A and 2C, respectively. One piece of blue transferprint (right) was also found in Unit 28/58 in Layer 2A. Taken together, these few artifacts suggest that not many personal items were taken to the site by the occupants and that the location may have been a place to focus on fishing and clamming.

It was only under the microscope and direct lighting that the outline of a figure was found. The figure was identified as a seated woman called a "US Seated Liberty Dime" (Yeoman 1995). This figure was used on the obverse side of a series of dimes. The figure changed slightly over time and these subtle changes make it possible to date this dime as minted from 1860 to 1891.

Several types of metal lids for bottles and jars were present in the Burton Acres Shell Midden, including pieces of zinc canning jar lids that began to be manufactured in 1858, crown caps for bottles that were first manufactured in 1892, and aluminum pull tabs beginning in 1962 (Firebaugh 1983). The crown caps and canning jar lids were found throughout all of the layers, while the pulltabs were found only in the uppermost layer.

One intriguing metal artifact is still unidentified (Figure 6.5). It was found in Layer 2A of Unit 29/58 and is possibly a decorated hinge. The metal is probably German silver which is comprised of 60% copper, 20% nickel and 20% zinc. The metal is stamped and ribbed. It is too thin (similar to an aluminum carbonated beverage can) to be a functional hinge or drawer pull. The decoration appears on one side only and is a very fine leaf pattern etched onto the surface.

Glass

As mentioned earlier, the glass assemblage is highly fragmented. However, there are a few diagnostic pieces that hint at what may have happened at this site. Unit 28/58 contained a partially reconstructible square alcohol (beverage) bottle. The bottle is clear with very uneven thickness from base to side and front panels. Several of the pieces have begun to exfoliate (sometimes called sick or patinated). They display an opalescent frosted appearance and small pieces of sodium carbonate have begun to flake off. This is due to the chemical decomposition within the glass when it comes in contact with moisture. Soda within the glass is brought to the surface and dries to form layers on the surface of the glass (Firebaugh 1983). This phenomenon is most common in bottles manufactured before 1920.

There were only three other diagnostic glass artifacts, two partial lip fragments and one three-piece mold fragment that could be reliably dated. A bottle lip (probably an alcohol bottle) from Layer 2A of Unit 28/58 was finished using an applied lipping tool that was used from 1870 to 1920. Another bottle lip, from a patent medicine or perfume bottle, was excavated from Layer 2A of Unit 29/58. The three piece mold (1870-1910) fragment was found in Layer 2D of Unit 22/58.

The absence of any dark olive green glass (prevalent before 1860), and the scarcity of any natural colored glass (aqua or cobalt blue), and the dates of the few

Figure 6.7 The top view of this metal button displays an NGW on its upper (outer) rim, which stands for the National Guard of Washington. Although this button was worn on the Guard's uniforms only from 1890 to 1902, we do not know if the button was dropped during this time. The button or the uniform jacket could have been recycled or saved as a valuable decorative object.

Figure 6.8 The underside view of the button from the National Guard of Washington uniform. The name of the company that made the button appears around the rim. This button was found in Unit 28/58 in Layer 2A, which is close to the surface. The button is 23 mm in diameter and weighs 4.8 grams.

diagnostic pieces suggest that the majority of the glass at Burton Acres Shell Midden dates to between 1860 and 1920. In the uppermost layer of the units, the presence of more contemporary glass, especially 7-Up green glass, is mixed with glass from the earlier period. But none of the more contemporary glass is found in the lower layers.

The assemblage contains no window glass. The two pieces of flat glass are either from mirrors or flat panels of bottles.

Many pieces of glass are burned, indicating that they were involved in a fire after breaking. In Unit 28/58, Layer 2A contained 17 fragments of burned glass, and Layer 2B contained a single fragment. In Unit 29/58, Layers 2A, 2B, and 2C each contained four burned pieces. Burned glass was found in no other layer. The distribution of burned glass matches that of other burned items suggesting that all kinds of objects were swept into a fire area together.

Ceramics

Ceramics are not well represented among the historic artifacts at Burton Acres Shell Midden. A few pieces of bricks and a minimal number of sherds from vessels are all that were recovered. Only one piece of

diagnostic pottery was found, a fragment of blue transferprint from Layer 2A of Unit 28/58 (Figure 6.6). Blue transferprint was manufactured between 1850 and 1910. Unfortunately, there is not enough of the decoration remaining to identify the pattern.

The undecorated ceramic assemblage consists of seven pieces of whiteware and five pieces of porcelain. Whiteware was manufactured in great numbers between 1820 and 1900, and continued to be manufactured in smaller numbers after 1900. Whiteware is found in large quantities in most historic sites in the Northwest because it was heavily used by the Hudson's Bay Company and was available at the Company stores (Chance and Chance 1976). The US military also used plain whiteware at their installations in the Northwest.

The greenish blue hue of the porcelain is indicative of Japanese porcelain, however the pieces are so small it is impossible to determine definitively. Porcelain, although not found in as great numbers as transferprint or whiteware, is also found at many other historic sites in the Puget Sound region.

Buttons

There were a total of seven buttons found at the site. Three of the buttons are ceramic and four are

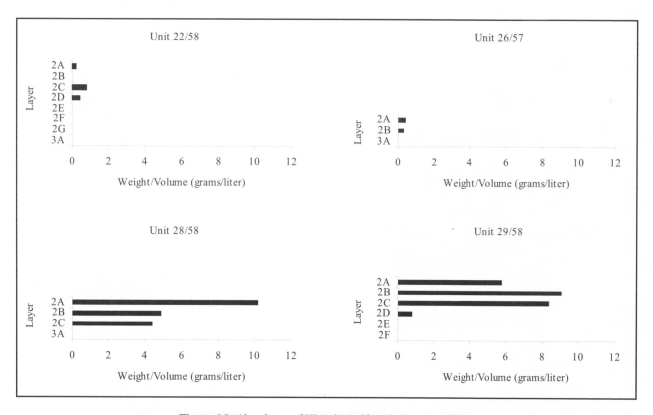

Figure 6.9 Abundance of Historic Artifacts by Layer and Unit

metal. The button in Figures 6.7 and 6.8 is from the dress blue uniform of the National Guard of Washington. The face of the button features an emblazoned eagle surrounded by stars with the letters "NGW" above the eagle (Johnson 1948). This type of uniform button was worn from 1890 to 1902. The other datable button is the Prosser button found in Layer 2C of Unit 29/58. Prosser type buttons were first manufactured in the 1840s. These buttons were generally plain white and used for utilitarian purposes. They can be identified by the small pinprick types holes on the back side surrounding the sew-through holes (Carley 1982).

Beads

The bead assemblage consists of only two beads. A white seed bead and a cobalt-blue seed bead. Seed beads were typical trade items and were used primarily as decoration on clothing because they were so small they could easily be sewn onto the fabric. The thin yet stiff sinew used by the Native Americans for sewing was much easier and more efficient to use than needle and thread (Oregon Archaeological Society 1965). Blue and white were the most popular colors for seed beads and were available in large quantities as trade items from the Hudson's Bay Company.

Personal Items

In addition to the buttons and beads, there are a only a few other artifacts that are typically considered to be for personal use. There was one insole of a leather shoe, a grommet/hook from a shoe, and a buckle from a piece of clothing.

RESULTS: QUANTITATIVE

This section consists of quantitative analyses of the historic artifacts.

Figure 6.9 shows the density of historic artifacts for all four units by layer. The weights of all historic artifacts were combined per layer and then divided by the volume in liters of each layer. This serves to standardize the weights and give a measure of density that can be compared horizontally and vertically. Historic artifacts appear in the highest concentrations in the uppermost layers of each unit. Significant increases in historic artifacts appear in Units 28/58 and 29/58. The absence of historic artifacts in the lower layers serves as a marker for changes that may have occurred in the activities at the site before and after contact.

Figure 6.10 shows that there are differences in the

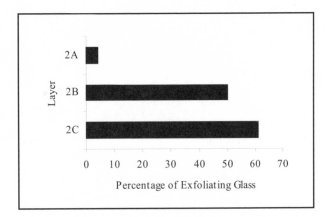

Figure 6.10 Distribution of Exfoliating Glass in Unit 29/58

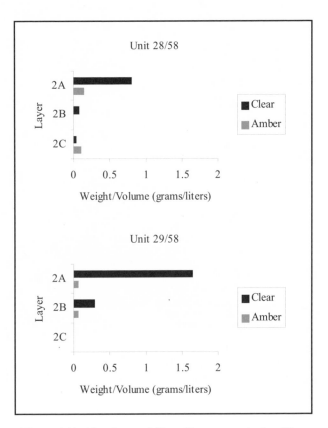

Figure 6.11 Abundance of Clear Glass versus Amber Glass

accumulation of glass artifacts between layers, and a closer examination of glass color and exfoliation shows that even subtle changes between layers are informative and can be detected. Many historical sites are excavated in large arbitrary units, because it has been the belief that the accumulation of artifacts has occurred over such a short amount of time that excavating by natural layers would be a waste of time and not provide any additional information. Exfoliation should be more prevalent in glass manufactured before 1920 and glass that has been exposed to moisture longer (such as burial in wet ground). The exfoliating glass in Unit 29/58 indicates that the layers do represent time. The higher concentrations of exfoliation in the lower layers could either be due to older glass, or glass that has been exposed to moisture longer. Both possibilities reflect the passage of time.

Another way to test the hypothesis that historical sites can be divided into smaller increments of time than previously believed is by examining the frequency of clear glass. Clear glass became popular for bottles and canning jars after 1916, especially after the change from using magnesium to using selenium to clear the glass. Magnesium in glass turns to an amethyst color through time, and selenium turns to a yellow (Firebaugh 1983). There are no amethyst-colored pieces of glass found at the site. There are, however, several pieces of clear glass that have a yellowish tinge indicating the use of selenium to clear the glass and a date of post-1916. Figure 6.11 clearly indicates that the frequency of clear glass is higher in the upper layers of Units 28/58 and 29/58. Therefore, the frequency of clear glass increases

through time in the assemblage at Burton Acres Shell Midden, and it is most likely selenium-cleared, post-1916, glass. Amber glass remained popular as a glass for beer bottles throughout this time. The higher numbers of clear glass found in the upper layers reflects both the expected increase of historic artifacts in general, and the change in the glass manufacturing industry, which began using more clear glass in the early part of the 20th century.

DISCUSSION

The historic artifacts that can be dated at the Burton Acres Shell Midden date from approximately 1860 to the present. There is a concentration of artifacts, dating between 1860 and 1920, in the uppermost layer, mixed with datable artifacts that post-date 1920.

Artifact manufacture dates can only be used as absolute dates to indicate the earliest possible time the artifact could have been deposited at the site. Artifacts often have long use-lives and are only discarded after they are no longer useful.

One obvious characteristic of the artifacts at the

Burton Acres Shell Midden is the lack of many artifacts that are typically associated as trade items, or items that date before 1860, the year that the Hudson's Bay Company's influence in the area diminished. The possible explanations for this are that the site was not occupied during the early Hudson's Bay years (late 1820s and early 1830s), or that the Native Americans who used the site did not use Euro-American artifacts. A third explanation is that they only occupied the site occasionally and for short time periods, therefore not depositing many artifacts before 1860. This trend supports the suggestion that the lives of the Native Americans in the Puget Sound region were not greatly altered by the Hudson's Bay Company. The greater influence may have come later with the missionaries, settlers, and US Government.

The lack of many personal items, structural items (particularly window glass), or household items, suggests that there was not a substantial or long-term structure at this site. The nails and other structural items found could indicate that a small dwelling existed at the site. However, the lack of any window glass is a good indication that a more permanent or substantial structure was not located here.

The presence of nails at this site could also be explained by fishing activities that could have taken place at the site. Herring in Puget Sound was harvested using a herring rake with bone teeth. After contact, these teeth were often replaced with square nails that were sharpened. The nails at this site, however, are so corroded that it is impossible to determine if they were modified for use in rakes.

The evidence from the historical artifacts suggests that this site was used as a seasonal camp that did not require permanent structures, and as a place where most of the inhabitants' personal and household items were removed at the end of the season. Finally, despite the introduction of European artifacts, primarily metal ones, after 1860, the activities at the site remained essentially unchanged from the prehistoric period.

REFERENCES

Carley, C.D.
1982 *HBC Kanaka Village/Vancouver Barracks 1977.* Reports in Highway Archaeology No. 8. Office of Public Archaeology, Institute for Environmental Studies, University of Washington, Seattle.

Chance, D.H.
1973 *Influences of the Hudson's Bay Company on the Native Cultures of the Colville District.* Memoir No. 2. Northwest Anthropological Research Notes. University of Idaho, Moscow.

Chance, D.H., and J.V. Chance
1976 *Kanaka Village/Vancouver Barracks.* Reports in Highway Archaeology No. 3. Office of Public Archaeology, Institute for Environmental Studies, University of Washington, Seattle.

Cole, D., and D. Darling
1990 History of the Early Period. In *Northwest Coast,* edited by W. Suttles, pp. 119-134. Handbook of North American Indians, vol. 7, W.C. Sturtevant, general editor, Smithsonian Institution, Washington, D.C.

Firebaugh, G.S.
1983 Archaeologist's Guide to the Historical Evolution of Glass Bottle Technology. *Southwestern Lore* 49(2):9-29.

Johnson, D.F.
1948 *Uniform buttons; American Armed Forces, 1784-1948* vol. 1. Century House, Watkins Glen, New York.

Johnson, M.M.
1943 *Ammunition; Its History, Development and Use, 1600-1943-.22 BB cap to 20 mm. shell.* W. Morrow, New York.

Nelson, L.H.
1968 Nail Chronology: As an Aid to Dating Old Buildings. *History News* 24(11).

Oregon Archaeological Society
1965 *Indian Trade Goods.* Oregon Archaeological Society, Portland.

Plew, M.G., K.M. Ames, and C.K. Fuhrman
1984 *Archaeological Excavations at Silver Bridge (10-BO-1), Southwest Idaho.* Archaeological Reports No. 12. Boise State University, Boise.

Serven, J. (editor)
1964 *The Collecting of Guns.* Bonanza Books, New York.

Steele, H.
1977 Euroamerican Artifacts in the Oregon Territory, 1829-60: A Comparative Survey. *Northwest Anthropological Research Notes* 11(2):174-183.

Suttles, W., and B. Lane
  1990 Southern Coast Salish. In *Northwest Coast*, edited by W. Suttles, pp. 485-502. Handbook of North American Indians. vol. 7, W.C. Sturtevant, general editor. Smithsonian Institution, Washington.

Yeoman, R.S. (editor)
  1995 *1996 Handbook of United States Coins*. 53rd ed. Western Publishing Company, Racine, Wisconsin.

7

Lithics

Timothy Allen

People living in Puget Sound in the past made their knives, spear and arrow tips, mauls, and adzes out of stone. The ingenious stone tools, called "lithics," meaning "made of stone," were shaped using methods that consisted of breaking (chipping) and grinding (pecking). The lithics found at Burton Acres Shell Midden are the tools people used in their daily lives. They are made of material from all over the Northwest, which indicates the complicated trade relationships needed to acquire these tools.

The excavations at the Burton Acres Shell Midden produced a relatively small lithic assemblage. A total of seventy-nine lithic artifacts were recovered, including both chipped and ground stone specimens. The majority of the chipped stone artifacts consist of debitage products. Debitage is the stone debris produced during the manufacture of chipped stone tools that is used by archaeologists to reconstruct the production of the tool. In addition to debitage, nine retouched specimens were recovered, including two identical unilaterally barbed triangular projectile points. Retouched means the object was shaped by chipping. Also, an analysis of lithic materials revealed a predominance of raw material not found locally (called exotics).

As the manufacture of stone implements is traditionally considered to be a "pre-contact" technology (Stewart 1996), this analysis examined change through time in the use of stone at this site. In addition, an analysis of imported lithic materials focuses on issues of regional trade.

METHODS

This analysis addresses 100% of the lithics recovered from the Burton Acres Shell Midden. Appendix D contains the data from the analysis of the lithic assemblage, as well as an explanation of the coding methods employed in the analysis.

RESULTS

Table 7.1 and Figure 7.1 display a summary of the lithic artifacts. Also included is the total volume of each layer that was excavated. The lithic assemblage from the Burton Acres Shell Midden is composed of seventy-nine specimens, the majority of which came from Units 22/58 and 29/58.

The assemblage consists predominantly of small fraction (1/8 inch) debitage, although there are a small number of shaped or retouched specimens (Figure 7.2; Table 7.2).

In this analysis, the lithic artifacts are addressed in terms of three groups: Material Types, Chipped Stone,

Chapter opening photo: Mary Parr (right) explains to schoolchildren how to recognize the difference between beach rock and stone debris produced during the manufacture of chipped stone tools.

Table 7.1 Summary of Lithic Artifacts by Unit and Layer

Unit	Layer	Volume (liters)	Chipped Stone						Groundstone			Total
			Shatter	Flake	Retouched Flake	Core	Biface	Projectile Point	Abrader	Modified Stone	Groundstone	
22/58	2A‡	36	1	2	-	-	-	-	-	-	-	3
	2B‡	142	-	8	-	1	-	-	-	-	-	9
	2C‡	60	-	2	-	-	-	-	-	-	-	2
	2D	58	-	1	1	-	-	-	-	-	-	2
	2E	48	-	8	-	-	-	-	-	-	-	8
	2F	22	1	-	-	-	-	-	-	-	-	1
	2G	16	-	-	-	-	-	-	-	-	-	-
	3A	24	1	-	-	-	-	-	-	-	-	1
	Unit Total		3	21	1	1	-	-	-	-	-	26
26/57	2A‡	102	3	3	-	-	-	-	-	-	-	6
	2B	36	-	-	-	-	-	-	-	-	-	-
	3A	8	-	-	-	-	-	-	-	-	-	-
	Unit Total		3	3	-	-	-	-	-	-	-	6
28/58	2A‡	132	-	2	-	-	1	-	-	-	-	3
	2B‡	156	-	2	-	-	-	-	-	-	-	2
	2C‡	112	-	-	-	-	-	-	-	1	-	1
	3A	32	-	1	-	-	-	-	-	-	-	1
	Unit Total		-	5	-	-	1	-	-	-	-	7
29/58	2A‡	154	-	12	-	-	-	-	-	-	-	12
	2B‡	82	-	4	-	-	-	-	-	-	2	6
	2C‡	164	1	6	-	-	-	-	-	-	-	7
	2D	96	-	4	1	-	-	4	-	-	-	9
	2E	126	-	2	-	-	1	-	1	1	-	5
	2F	80	-	-	1	-	-	-	-	-	-	1
	Unit Total		1	28	2	-	1	4	1	1	2	40
Total		1686	7	57	3	1	2	4	1	2	2	79

‡ = layers thought to date to post-contact with Euro-Americans; all other layers date to pre-contact (see Table 5.1).

and Groundstone. Seventy-four of the lithic artifacts (94%) are chipped stone, while only five artifacts (6%) are groundstone. In addition, the chipped stone artifacts are further divided into Debitage Products (flakes, shatter), and Cores and Retouched Specimens (cores, retouched flakes, bifaces, and projectile points).

Materials

Figure 7.3 presents a summary of lithic raw materials (rock type of each piece of debitage or tool) recovered from the Burton Acres Shell Midden. The lithic assemblage is dominated by jasperoid and dacite specimens. Other materials identified in the assemblage include vitric tuff, chalcedony, jasper, petrified wood, and a fine-grained igneous material, possibly trachyte (Edward F. Bakewell, personal communication 1997).

Significantly, the only chipped stone material recovered from Burton Acres Shell Midden that occurs naturally on the Puget Sound region is dacite. Dacite is

a volcanic material (Bakewell 1993, 1996), and is common in western Washington. Jasperoid is "a carbonate mud rock" (Hayden et al. 1996:349), and does not occur naturally on the west side of the Cascade Range. Vitric tuff is a volcanic material (Hayden et al. 1996; Bakewell 1993), found in southwest British Columbia (Edward F. Bakewell, personal communication 1997), and recently observed in the Cedar and Snoqualmie River drainages (Maury Morgenstein, personal communication 1999). Trachyte also occurs in southwestern British Columbia (Edward F. Bakewell, personal communication 1997). Chalcedony occurs in eastern Washington and southwest British Columbia (Edward F. Bakewell, personal communication 1997). Petrified wood is found in eastern Washington. The two most abundant materials, dacite and jasperoid, are the only ones recovered from both the shallowest and the deepest excavated layers.

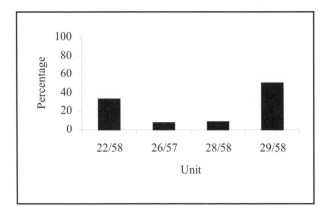

Figure 7.1 Number of lithic artifacts found in each excavation unit.

A useful way to consider the lithic materials is in terms of "local" and "non-local" (Figure 7.4). For this analysis vitric tuff is considered non-local, although this designation has been called into question and may change with future research. The assemblage is characterized by a predominance of "non-local" materials intermingled throughout the excavated layers. Neither type of material was exploited to the complete exclusion of the other during any portion of the temporal sequence. Interestingly, only two of the assemblage's retouched specimens, a retouched flake (Cat. # 29582F0023001), and a biface (Cat. # 28582A0353400) are made of a "local" material. Considering the small size of the sample and the potential problem with vitric tuff, it is unknown whether this observation is significant.

CHIPPED STONE

The chipped stone artifacts comprise the majority of the assemblage, and consist of debitage (including seven pieces of shatter and fifty-seven flakes), one core, and retouched specimens (three retouched flakes, two bifaces, and four projectile points).

Debitage

Flakes are the most abundant lithic artifacts recovered from Burton Acres Shell Midden, and they occur throughout most of the excavated layers (Figure 7.5). Flakes are absent, however, from the lower layers of Units 22/58, 26/57, and 29/58. As the overall lithic density at this site is so low, it is uncertain whether the absence of flakes from some of the lowest layers is significant. The majority of the flakes were recovered from Units 22/58 and 29/58.

Table 7.2 Lithic Artifacts by Screen Size

Unit	Layer	Volume	Screen Size				Total
			1"	1/2"	1/4"	1/8"	
22/58	2A‡	36	1	-	-	2	3
	2B‡	142	-	1	1	7	9
	2C‡	60	-	-	-	2	2
	2D	58	-	1	-	1	2
	2E	48	-	-	1	7	8
	2F	22	-	-	1	-	1
	2G	16	-	-	-	-	-
	3A	24	-	-	-	1	1
Unit Total			1	2	3	20	26
26/57	2A‡	102	-	1	1	4	6
	2B	36	-	-	-	-	-
	3A	8	-	-	-	-	-
Unit Total			-	1	1	4	6
28/58	2A‡	132	-	-	1	2	3
	2B‡	156	-	-	-	2	2
	2C‡	112	1	-	-	-	1
	3A	32	-	-	-	1	1
Unit Total			1	-	1	5	7
29/58	2A‡	154	-	-	3	9	12
	2B‡	82	2	-	1	3	6
	2C‡	164	-	-	4	3	7
	2D	96	1	1	5	2	9
	2E	126	3	-	1	1	5
	2F	80	1	-	-	-	1
Unit Total			7	1	14	18	40
Total		1686	9	4	19	47	79

‡ = layers thought to date to post-contact with Euro-Americans; all other layers date to pre-contact (see Table 5.1).

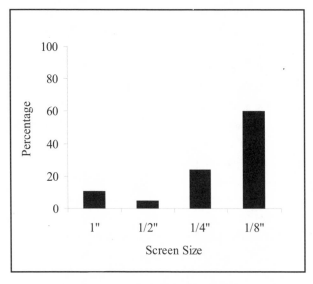

Figure 7.2 Number of lithics found in each screen size.

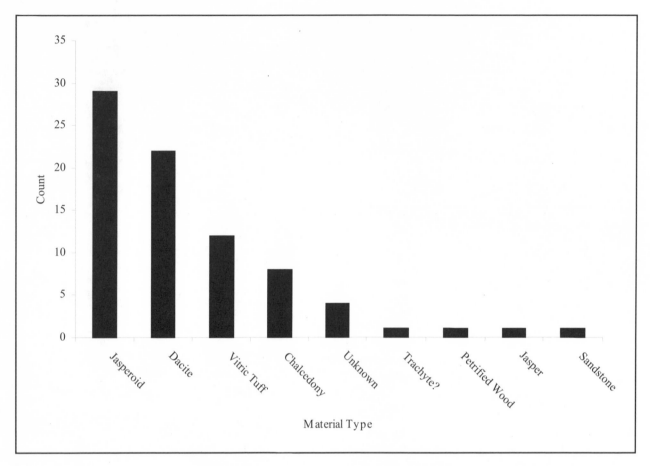

Figure 7.3 Number of lithics composed of each rock type.

Of the 57 flakes recovered from the site, 74% were drawn from the 1/8 inch mesh screen (Figure 7.6). Shatter was also primarily recovered from the 1/8 inch mesh screen. Only one flake was characterized as "secondary" (retaining 25%-50% dorsal cortex), while no "primary" (retaining 50%-100% dorsal cortex) flakes were identified. All other debitage products were "tertiary" (retaining 0%-25% dorsal cortex), exhibiting no dorsal cortex. The diminutive size of both the flakes and the assemblage, and the presence of only one "secondary" flake, indicate that significant chipping of stone material did not occur at the Burton Acres Shell Midden. Rather, these observations suggest only minor tool maintenance occurring here.

Cores

The only core recovered from the site is a dacite pebble with a single flake removed from the weathered exterior. This artifact was recovered from Layer 2B of Unit 22/58. The recovery of this single core from this site further indicates that chipped stone tools were not manufactured here.

Retouched Specimens

Retouched specimens (a flake possessing secondary retouch on one or more of its edges) comprise 12% of the chipped stone artifacts, and include two bifaces, three retouched flakes, two complete projectile points, and two incomplete projectile points. The majority of the retouched specimens were recovered from Unit 29/58: five from Layer 2D, one from Layer 2E, and one from Layer 2F. The other two were recovered from Unit 22/58, Layer 2D, and Unit 28/58, Layer 2A. The retouched specimens represent a variety of materials, including chalcedony, dacite, vitric tuff (possible Duck Meadow chert), petrified wood, jasperoid, and a fine grained igneous material, possibly trachyte (Edward F. Bakewell, personal communication 1997).

Descriptions of projectile points and bifaces employ a fairly standard set of terms (Justice 1987; Lohse 1985), while descriptions of retouched flakes employ a wide variety of attributes. Seven characteristics of retouch are used here (Inizan et al. 1992): *position, localization, distribution, delineation, extent, angle,* and *morphology.*

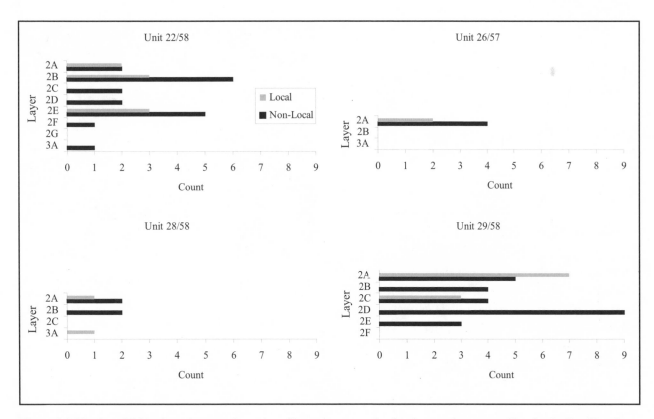

Figure 7.4 Number of lithics in each excavation unit and layer that are made of rock types that are considered to be "local" or "non-local" sources.

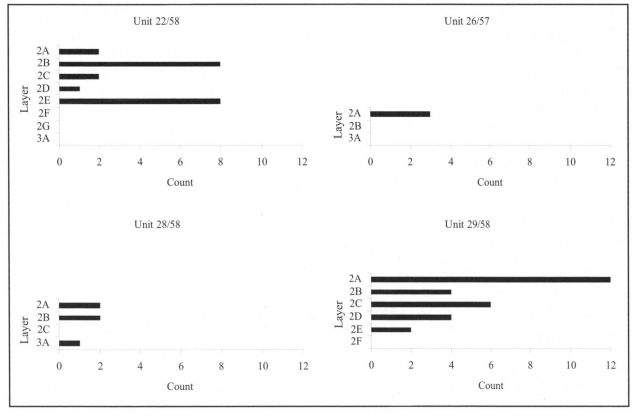

Figure 7.5 Number of flakes (one kind of lithic) found in each excavation unit and layer.

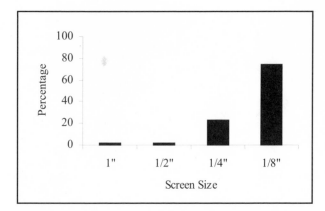

Figure 7.6 Number of flakes found in each screen size.

Position refers to the location of small flakes (called removals) relative to the faces of an artifact. In this way, "inverse" refers to removals located on the ventral face, "obverse" to those on the dorsal face, and "bifacial" to removals on both faces of a margin. *Localization* indicates the margin upon which removals occur, relative to a specific orientation. Flakes are oriented with the ventral face down, and the proximal end towards the analyst. *Distribution* describes the continuity of removals along the length of an edge. Removals are thereby described as continuous, discontinuous, or partial. *Delineation* describes the outline of an edge formed by a series of removals: straight, convex, concave, or irregular. *Extent* describes the invasiveness of removals on the faces of an object. In this way, removals are described as short, long, invasive, or covering. *Angle* indicates the "angle formed by removals relative to the face from which they are taken" (Inizan et al. 1992:75), and are low (@ 10º), semi-abrupt (@ 45º), and abrupt (@ 90º). Finally, the *morphology* of removals is described in four terms: scaled, stepped, parallel, and sub-parallel.

Retouched Flakes

Three retouched flakes were recovered: two from Unit 29/58, and one from Unit 22/58. Two of these specimens are jasperoid, and the third is dacite. The dacite retouched flake (Cat. # 29582F0023001) is exceptionally large in comparison to the other chipped stone specimens, weighing 240g. The chipped stone artifact that is the nearest to it in weight is a 13.3g flake (Figure 7.7). The other two retouched flakes weigh 2.7g and 1.2g, respectively.

The dacite specimen exhibits inverse, right, continuous, short, low, stepped retouch which produces

Figure 7.7 This flake is made from a fine-grained volcanic rock, most likely dacite. It is considered a retouched flake because a smaller, secondary flake has been removed from one or more of edges – thus "retouched." This particular specimen was the largest lithic found at the site, weighing 240g (about the size of your fist). Other sites in the Northwest contain larger numbers of retouched flakes than the Burton Acres Shell Midden, suggesting that prehistoric people did not make stone tools here.

a convex margin. One of the jasperoid specimens (Cat. # 22582D0263200) possesses retouch on its proximal, distal, and right margins. The proximal retouch is inverse, partial, short, low, and subparallel, and creates a straight edge. The distal retouch is inverse, continuous, long, low, and parallel, and produces a convex outline. The right retouch is inverse, partial, long, low, and subparallel, and produces a straight margin. The other jasperoid specimen (Cat. # 29582D0023001) possesses retouch on its distal and right margins. The right retouch is bifacial, continuous, long, low, and subparallel, and creates an irregular margin. The distal retouch is obverse, continuous, short, abrupt, and subparallel, and also produces an irregular margin.

Bifaces

Two bifaces were recovered: one from Unit 28/58 (Cat. # 28582A0353400), and one from Unit 29/58 (Cat. # 29582E0113001). One specimen is vitric tuff (possibly Duck Meadow chert) (Edward F. Bakewell, personal communication 1997), and the other is dacite. Both bifaces are incomplete specimens, and both are potentially fragments of projectile points. Due to a lack of distinguishing features, however, these artifacts are described only as being bifacially retouched. The vitric tuff specimen has been heat-altered, and is possibly the base of a projectile point (Figure 7.8). This specimen

Figure 7.8 This lithic is retouched on both sides (faces) and, therefore, is called a "biface." It was probably the base of a projectile point, but because it is broken we cannot know for sure. This biface is made of vitric tuff, a rock created during fast-cooling volcanic activity. Vitric tuff is found throughout the Cascade Mountains.

Figure 7.9 This retouched flake has been worn by water and probably laid on the beach in the wave zone before it was thrown into a pile of shells, rock, and charcoal. It is so small that we cannot tell what part of a tool it represents.

Figure 7.10 This small, broken projectile point has notches on the side of its base to facilitate attachment to an arrow shaft or handle. The rock type is chalcedony, a type of chert found on the eastern side of Cascades. The tip of this projectile broke off, after which it was retouched. Retouching broken points indicates that people valued this raw material and used it until it was too small to function adequately.

exhibits covering bifacial retouch, and significant basal thinning. The base is straight, and the blade margins appear to be excurvate. The dacite specimen is considerably worn, and appears to exhibit covering bifacial retouch (Figure 7.9). This specimen's three intact margins are straight.

Projectile Points

All four of the projectile points came from Unit 29/58, Layer 2D. The two incomplete projectile points are made of chalcedony. One (Cat. # 29582D0013404)

is side-notched, and bears evidence of being broken and subsequently retouched (Figures 7.10 and 7.13a). The other incomplete projectile point (Cat. # 29582D0013403) has a concave base and slightly excurvate blade margins (Figures 7.11 and 7.13d).

The two complete projectile points are the most interesting lithic artifacts that were recovered from the site. These nearly identical specimens may be described as unilaterally barbed triangular points. Both of these points possess excurvate blade margins. One (Cat. # 29582D0013402) is made of an igneous material, possibly trachyte (Edward F. Bakewell, personal communication 1997) (Figures 7.12a and 7.13b); the other (Cat. # 29582D0013201) is made of petrified wood (Figures 7.12b and 7.13c). Neither material occurs naturally in the Puget Sound region. E.F. Bakewell

Figure 7.11 The tip of this small, chalcedony projectile point has been broken off, leaving only the base. Chalcedony is a type of chert that was highly prized for its excellent flaking properties. The tips of projectile points are often missing because they break on impact.

Figure 7.12 Two, nearly identically shaped projectile points were found together in the deepest layer of Unit 29/58, the unit closest to the water. These points are unusual, not only because they are similarly shaped, but also because they are asymmetrical. Each one has a longer tang (here oriented with the longer tang on the left) on one side. The specimen on the left is made of fine-grained volcanic rock, probably trachyte. The one on the right is made of petrified wood. These points were probably attached to arrows.

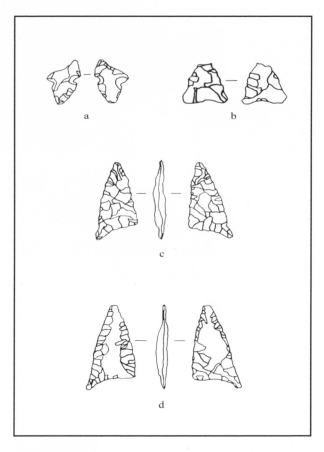

Figure 7.13 Projectile points recovered from the Burton Acres Shell Midden: (a) broken chalcedony projectile point, (b) broken chalcedony projectile point, (c) trachyte projectile point, and (d) petrified wood projectile point. These drawings depict the relative size of the four projectile points found at the site (drawing by T. Allen).

(personal communication 1997) notes that trachyte resembling the specimen from Burton Acres Shell Midden occurs in southwestern British Columbia, and that petrified wood is found east of the Cascade Range. This style of projectile point is not common in the Puget Sound area, or elsewhere in the Pacific Northwest.

A similar, unilaterally barbed point base was recovered from the Mule Spring Site (45KI435) (Miss and Nelson 1995:38), however the specimen is too fragmentary to determine if the complete artifact would have resembled those recovered from Burton Acres Shell Midden. Asymmetrical (acute, or non-equilateral) triangular points have been found at many sites in the Puget Sound area, including British Camp (45SJ24) (Stein 1992), Old Man House (45KP2) (Snyder 1956), and Tualdad Altu (45KI59) (Chatters 1988:80), and exhibit a general outline that is relatively similar to the Burton Acres Shell Midden points. None of these

Figure 7.14 This abrader was used to sharpen wooden, bone, metal, or stone objects in the same manner we use whetstones today to sharpen knives. The surface viewed here is very smooth to the touch, indicating a great deal of use.

Figure 7.15 This large rounded rock is battered on one end and is, therefore, called a "hammerstone." Such battering is produced when the stone is used in the same way we would use a hammer.

specimens, however, possess the distinct barb exhibited by the Burton Acres Shell Midden points.

GROUNDSTONE

The groundstone artifacts consist of one sandstone abrader, two pecked/polished stones, and two miscellaneous specimens. The abrader and one pecked/polished stone and both miscellaneous artifacts, were recovered from Unit 29/58, while the remaining pecked/polished stone came from Unit 28/58.

Abrader

The abrader is characterized by a thin, rectangular cross-section, and evidence of smoothing through abrasion on both faces (Figure 7.14). This artifact was recovered from Layer 2E of Unit 29/58.

Pecked/Polished Stone

One of the pecked/polished stones exhibits pecking on one end, and may be described as a "hammerstone" (Figure 7.15). The other stone exhibits a fine gloss on both ends. Both of these objects are water-worn cobbles. In this way, these objects are distinctive, because such cobbles are rare on the beach occupied by the Burton Acres Shell Midden.

Miscellaneous Specimens

The two miscellaneous specimens were recovered from Layer 2B of Unit 29/58. One (Cat. # 29582B0193100) is a large, nearly oval stone that exhibits evidence of being ground on one of its surfaces. The other object (Cat. # 29582B0303100) is consider-

ably smaller and also exhibits evidence of grinding on one surface. Both artifacts possess single, small, rust-colored stains, indicating that iron objects oxidized while in contact with these specimens.

DISCUSSION AND CONCLUSIONS

Excavations at the Burton Acres Shell Midden recovered a total of 79 lithic artifacts. The assemblage consists primarily of chipped stone artifacts (94%), and contains only five (6%) groundstone objects. The assemblage is characterized by a predominance of small fraction (1/8 inch) debitage, and a relatively small number of shaped or retouched specimens. Indeed, 60% of the entire assemblage was drawn from 1/8 inch mesh screen.

The size and contents of the lithic assemblage indicate that chipped stone tools were not manufactured at this site. Although flakes are the most abundant lithic artifacts recovered from the site, they compose a very small sample. In addition, the vast majority of the site's flakes were recovered from the 1/8 inch screen, and only one specimen exhibited noticeable dorsal cortex. Also, only one core was recovered from the site. These observations suggest minor tool maintenance rather than tool manufacture.

Lithic artifacts were recovered throughout the excavated layers and are actually less abundant in the lower "pre-contact" layers. This is contrary to the hypothesis that lithic tool manufacture was only a "pre-contact" technology that was replaced by metal technology. The consistently low density of stone artifacts indicates that the manufacture of stone implements was never particularly intense at this location. It is notable, however, that retouched chipped stone objects (ones where secondary flakes were removed from the edges) are located predominantly in the lower, "pre-contact" layers. Perhaps people occupying the site were using the lithics more intensively at that time.

The lithic materials recovered from the site represent both "local" and "non-local" varieties. Jasperoid, a "non-local" material, is the most abundant material recovered from the site, while dacite, a material that occurs naturally in the Puget Sound region, is the second most abundant. In fact, each of the other five materials represented by chipped stone artifacts (vitric tuff, chalcedony, petrified wood, jasper, and a fine grained igneous material, possibly trachyte) do not occur naturally in the Puget Sound region. Although "non-local" materials are definitely the most abundant, the presence of dacite throughout the temporal sequence suggests that at no point were exotics exploited to the complete exclusion of "local" material. Interestingly, however, only two of the nine retouched specimens, a retouched flake and a biface, are made of dacite. It is perhaps significant that the two identical unilaterally barbed triangular projectile points that are anomalous in the Puget Sound region are made of exotic materials. Both exotic and "local" materials are present before and after the period of Euro-American contact. It appears evident that use of lithic materials at the Burton Acres Shell Midden, while never intense, continued after Euro-American contact.

REFERENCES

Bakewell, E.F.

1993   Shades of Gray: Lithic Variation in Pseudobasaltic Debitage. *Archaeology in Washington* 5:23-32.

1996   Petrographic and Geochemical Source-Modeling of Volcanic Lithics from Archaeological Contexts: A Case Study from British Camp, San Juan Island, Washington. *Geoarchaeology: An International Journal* 11(2):119-140.

Chatters, J.C.

1988   *Tualdad Altu (45KI59): A 4th Century Village on the Black River, King County, Washington*. First City Equities, Seattle.

Hayden, B., E. Bakewell, and R. Gargett

1996   The World's Longest Lived Corporate Group: Lithic Analysis Reveals Prehistoric Social Organization near Lillooet, British Columbia. *American Antiquity* 61(2):341-356.

Inizan, M.L., H. Roche, and J. Tixier.

1992   *Technology of Knapped Stone*. Prehistoire de la Pierre Taillee, Tome 3. CREP, Meudon, France.

Justice, N.D.

1987   *Stone Age Spear and Arrow Points of the Midcontinental and Eastern United States*. Indiana University Press, Bloomington.

Lohse, E.S.

1985   Rufus Woods Lake Projectile Point Chronology. In *Summary of Results, Chief Joseph Dam Cultural Resources Projects, Washington*, edited by S.K. Campbell, pp. 317-364. Submitted to United States Army Corps of Engineers, Seattle District, Contract No. DAC W67 78 C 0106. Prepared by Office of Public Archaeology, Institute for Environmental Studies, University of Washington, Seattle.

Miss, C.J., and M.A. Nelson

1995   *Data Recovery at the Mule Spring Site, 45-KI-435, King County, Washington*. Northwest Archaeological Associates, Inc., Seattle.

Snyder, W.

1956   Archaeological sampling at "Old Man House" on Puget Sound. *Research Studies of the State College of Washington* 24:17-37.

Stein, J.K. (editor)

1992   *Deciphering a Shell Midden*. Academic Press, New York.

Stewart, H.

1996   *Stone, Bone, Antler and Shell: Artifacts of the Northwest Coast*. Douglas and MacIntyre, Ltd., Vancouver, British Columbia.

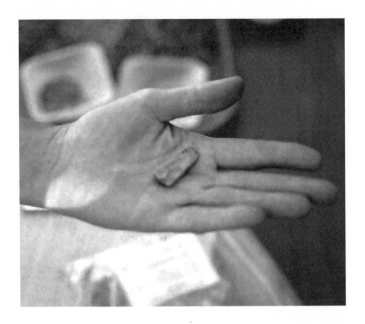

8

Bone and Antler Tools

Laura S. Phillips

For thousands of years, bone tools were an important part of the tool kit used by Native Americans in the Puget Sound region. It is not surprising, then, that tools made from bone and antler are found at the Burton Acres Shell Midden. An unexpectedly large number of tools were recovered in part because the abundant shell in the site changed the soil chemistry to favor bone preservation. The recovered adze parts, awls, and points shed light on woodworking, fishing and textile activities.

An analysis of modified bone often provides insights into the cultural activities performed at a site. Fourteen modified bone and antler tools were found in the four excavation units at the Burton Acres Shell Midden. Other shell midden sites in the region do not have as high a ratio of bone and antler tools to matrix. The relatively high abundance of bone and antler tools suggests two things about the midden: 1) people concentrated their activity in the small area sampled in our excavation; and 2) people used bone tools to work wood and fish at the site.

METHODOLOGY

This chapter reports the analysis of bone and antler tools found at the Burton Acres Shell Midden. Bones cut during butchering are not considered tools and are discussed in Chapter 9.

Just over half of the tools were found in situ, while the others were recovered in the 1 inch, 1/2 inch, 1/4 inch, and 1/8 inch mesh screens.

CLASSIFICATION

Table 8.1 is a summary of the bone tools found at the site. Bone tools are assigned a proposed function based on a limited classification system following zoological taxonomy (Larson and Lewarch 1995; Stewart 1996). Classification characteristics included shape, cross-section, type of modification, and location of modification. The descriptive summary below is discussed by artifact function, and details the characteristics used to determine the proposed function.

Adze Haft

In the past, adze blades in the Northwest Coast were usually made of stone, and attached to a handle (hafted) made of either bone or antler (Stewart 1996). Broken handles and used-up blades are often found separately in sites. At Burton Acres Shell Midden only the handles were found. It is likely adze blades were used in these handles (Stewart 1996).

Portions of two antler adze hafts and a possible third haft were found in situ in Unit 29/58, Layers 2E

Chapter opening photo: This bone chisel was discovered in an excavation unit where few other artifacts and almost no shell was found. Here it is being shown to visitors at the site.

Table 8.1 Bone and Antler Tools

Catalog Number*	Material	Description	Length (cm)	Width (cm)	Thickness (cm)	Weight (g)
22582E0162001	Bone	Double-bevel chisel	3.4	1.3	0.7	3.5
22582E0182001	Bone	Toggling harpoon valve	4.6	0.8	0.5	1.0
26572A0142401	Bone	Unipoint	2.7	0.2	0.3	0.1
29582A0462401	Bone	Modified bird rib	3.7	0.9	0.3	0.1
29582D0062801	Bone	Point tip	1.0	0.3	0.1	0.1
29582E0012001	Antler	Adze handle fragment	13.0	9.0	6.2	210.1
29582E0042401	Bone	Modified mammal limb bone	4.9	0.6	0.5	2.0
29582E0072001	Bone	Modified mammal bone	12.1	1.3	0.8	11.6
29582E0082001	Antler	Adze handle fragment	6.4	4.8	2.8	30.2
29582E0122001	Bone	Awl, needle-pointed	7.0	1.0	0.5	2.4
29582E0152001	Bone	Awl, needle-pointed	9.3	0.9	0.4	4.1
29582F0022002	Antler	Adze handle fragment	11.0	4.9	2.0	54.4
29582F0022100	Bone	Modified mammal limb bone	11.4	1.8	0.8	9.5
29582F0062001	Bone	Flaked mammal bone	4.3	0.8	0.7	2.2

*Catalog Number is a unique number containing provenience information. For example, 22582E0162001 is Unit 22/58, Layer 2E, Bucket Number 016, Material Code 2, Size Code 0, and Field Specimen Number 01 (see Tables 4.1 and 4.2).

and 2F. The two hafts (Cat. # 29582E0012001 [Figures 8.1 and 8.2] and Cat. # 29582F0022002 [Figure 8.3]) are faceted on both ends of the beam, and the cancellous tissue has been partially removed from one end. Both are split and broken longitudinally, suggesting the break occurred when a chisel blade was inserted or, perhaps, sometime during tool use. Curiously, a portion of the beam of Cat. # 29582E0012001 is a branch junction. This branch has been cut off, and the break has been cut and facetted as well. A portion of the cancellous tissue has been removed from this branch, perhaps for a handle (Figures 8.1 and 8.2).

The third antler beam (Cat. # 29582E0082001) is faceted at one end; the other end has broken off. This antler beam, although modified in the same style as the two hafts, is too fragmented to ascertain whether it was used as a haft.

Awl

Awls are bone points often made from splinters of long bone and are thought to have been used to perforated various organic materials. Two awls were found in Unit 29/58, Layer 2E. One (Cat. # 29582E0152001) is a long bone fragment that has been split perpendicular to the shaft at the distal end. It is ground, tapers to a polished point, and is ovoid in cross-section. The other awl (Cat. # 29582E0122001) is a splinter of a long bone. The splinter is triangular in cross-section, and two

of the three sides exhibit remnants of the marrow cavity. The third side, the exterior of the bone, has been ground. One end of the long bone has been ground and polished to a rounded tip.

Chisel

Ethnographically, a chisel was a bone blade attached to a handle and used for woodworking. A double-beveled chisel (Cat. # 22582E0162001) was recovered from Unit 22/58, Layer 2E (Figure 8.4). Although not as strong as a nephrite chisel, a bone chisel could have been used for more detailed woodworking. The small size of the chisel suggests that it was hafted.

Composite Toggling Harpoon Valve

The toggling harpoon, a piece of fishing gear important to Northwest people, is made of multiple parts including a point, two valves, an attachment rope, and a binding mechanism (Stewart 1996:110). A single grooved valve from a composite toggling harpoon (Figure 8.5) was found in Unit 22/58, Layer 2E at the same depth (39cm below surface) as the chisel. The valve (Cat. # 22582E0182001) has the characteristic groove and rounded sockets that Hilary Stewart (1996:110) suggests were used for salmon procurement. Campbell (1981:289) describes two forms of valves. Both have a socket on the spur; Campbell's Type I has a socket on the body that is similar to that on the spur,

Figure 8.1 This fragment of antler (to the right of the arrow) was so clearly modified that it was recognized as an artifact immediately upon uncovering it. The ends of the antler have been removed and beveled, most likely to make an adze handle.

Figure 8.3 This antler fragment is broken along the long dimension facing the photographer. The split probably occurred when a chisel blade was inserted into one end of the handle, or else when the chisel was used, and too much pressure fractured the handle. Clearly, people threw this away when it was too broken to be useful.

Figure 8.2 This view of the antler adze handle, seen in Figure 8.1, shows the faceted ends.

while Type II has a flat surface and no socket. In contrast, the one found at Burton Acres Shell Midden fits neither form. It has a foreshaft socket like the Campbell's Type I and II, but the point socket is a slight depression.

Point

Bone points, thin splinters of sharpened bone, are often found in Northwest Coast sites, and their function is difficult to assign. Ethnographically, they were used

Figure 8.4 This bone has been sharpened on both edges of one end and most likely was made to be inserted into a handle. Bone chisels were used for carving wood and making wooden implements. This one was not broken, so perhaps it was dropped accidentally and lost.

Figure 8.5 This grooved piece of bone is part of a complex fishing gear called a toggling harpoon. It is in perfect condition, not broken or cracked. Its maker would have needed two such grooved valves to press on both sides of a point and bind with cordage. Only one valve was found at the site, but its presence suggests that people were carrying all sorts of fishing gear here and likely fishing for salmon.

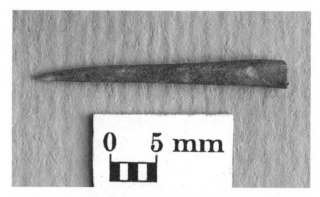

Figure 8.6 This bone point has been sharpened on only one end (a unipoint) and is believed to have been used as a barb in fishing gear. Barbs were attached to all kinds of hooks to prevent a fish from sliding off after taking the bait. Bone points are made from splinters of mammal long bone, shaped and sharpened on stone abraders

as needles, harpoon points, and parts of herring rakes. A herring rake is a long pole for stabbing herring. It consists of wood embedded with barbs along one edge extending approximately a third of the pole's length (see Stewart 1996:102).

Two bone points were recovered. One, a unipoint (Cat. # 26572A0142401), was found in Unit 26/57, Layer 2A (Figure 8.6). Its relative narrowness and diameter-to-length ratio (1:9) indicates that it may have been used as a barb or needle. A variety of uses for barbs has been suggested, from hafted barbs on fish hooks to barbs on herring rakes (Stewart 1996; Roll 1974). It is probably not a harpoon point because of its small size and narrowness.

The other point (Cat. # 29582D0062801) is fragmentary, and comprises the tip only. Recovered from Unit 29/58, Layer 2D, this tip is triangular in cross-section, and ground along one edge. Due to the fragmentary nature of this artifact no function is assigned.

Worked Bone

Four mammal bones and one bird bone were modified, but are not formed tools. They show evidence of working, but are not shaped into a tool for which the function is recognizable. Three bones have been ground and beveled. The other two are long bone splinters that appear to have been flaked.

RESULTS

Abundance

Despite the small size of the excavation units, a comparatively large number of bone tools were found,

approximately nine per cubic meter. This is at least four times the number of bone tools per cubic meter reported recovered from other sites in the Puget Sound region (Blukis Onat 1987; Bryan 1963; Campbell 1981; Chatters 1988; Larson 1996; Larson and Lewarch 1995; Mattson 1971; Wessen 1989).

In general, shell middens lack large quantities of formed tools. The larger number of bone tools from Burton Acres Shell Midden were found in the lower layers of the site, and may indicate a heavy-use or discard area where tools would accumulate in greater than average concentrations (Figure 8.7). Over 80% of the bone tools were recovered from the lowest layers of Unit 29/58. Interestingly, nearly all of the retouched lithics were also found in these layers. The volume of shell in these lower layers is also significantly higher than in the rest of the site.

One likely factor contributing to the abundance of bone tools at this site is excellent bone preservation. Nearly all of the bone tools were found in the lower layers of Unit 29/58, where ground water inundates the site as the tide fluctuates. The tools seem to have been thrown into a wet intertidal zone, which protected them from weathering. In addition, the large quantity of shells increased the likelihood of preservation due to the basic pH of calcium carbonate that comprises shell.

Activity Shifts

Only one formed tool, a unipoint, was found in the upper layers; all of the woodworking tools were found in the lowers layers. While preservation is

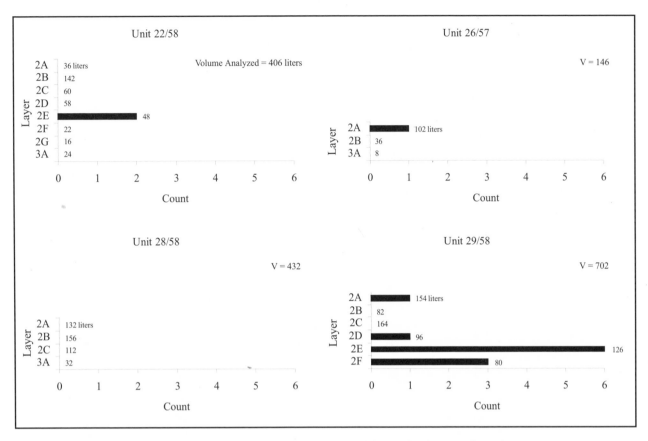

Figure 8.7 Number of bone tools found in each layer of each excavation unit.

surely a factor in this high concentration of bone tools in the layer, the relative abundance of wood-working tools suggest an activity change. One possibility is that the location of woodworking shifted to another part of the site (an area not excavated). Another possibility is that of European impact. If the lower layers represent deposition of objects during a time before European contact, then perhaps the apparent lower number of formed tools in the upper layers indicates a change in activity focus after the introduction of European artifacts (i.e. away from using bone tools for woodworking). The sample size of bone and antler tools is too small to test statistically the significance of this apparent shift in tool use. However, the distribution of bone and antler tools considered with the faunal and floral evidence suggests the shift was affected by objects introduced by Euro-Americans, including metal.

CONCLUSIONS

The number of bone tools found at the Burton Acres Shell Midden is small, but provides insights into the possible activities at the site. The presence of multiple

adze hafts as well as a chisel suggests that people were working wood at the site. Fishing tools were also recovered, including the composite toggling harpoon valve and points. The awls could have been used for piercing hide or bark, or for basketry weaving.

The bone tools found at the Burton Acres Shell Midden are not unique, as they have been found at many of the sites in Puget Sound. They are representative of a tool technology shared throughout the region.

REFERENCES

Blukis Onat, A.

1987   Comparative Modified Artifact Analysis. In *The Duwamish No. 1 Site: 1986 Data Recovery*, edited by URS Corporation Seattle and BOAS, Inc., pp. 7.1-7.31. Submitted to Municipality of Metro-politan Seattle (METRO), Contract No. CW/F2-82, Task 48.08. Prepared by BOAS, Inc., Seattle.

Bryan, A.L.

1963   *An Archaeological Survey of Northern Puget Sound*. Occasional Papers of the Idaho State University Museum No.11, Pocatello.

Campbell, S.K. (editor)

1981   *The Duwamish No. 1 Site: A Lower Puget Sound Shell Midden*. Research Report No. 1. Office of Public Archaeology, Institute for Environmental Studies, University of Washington, Seattle.

Chatters, J.C.

1988   *Tualdad Altu (45KI59): A 4th Century Village on the Black River, King County, Washington*. First City Equities, Seattle.

Larson, L.L. (editor)

1996   *King County Department of Natural Resources Water Pollution Control Division Alki Transfer/ CSO Project: Allentown Site (45KI431) and White Lake Site (45KI438 and 45KI438A) Data Recovery*. Submitted to HDR Engineering, Inc., Bellevue, Washington. Prepared for King County Department of Natural Resources. Prepared by Larson Anthropological/Archaeological Services, Seattle.

Larson, L.L., and D.E. Lewarch, (editors)

1995   *The Archaeology of West Point, Seattle, Washington 4,000 Years of Hunter-Fisher-Gatherer Land Use in Southern Puget Sound*. Submitted to CH2M Hill, Bellevue, Washington. Prepared for King County Department of Metropolitan Services. Prepared by Larson Anthropological/Archaeological Services, Seattle.

Mattson, J.L.

1971   A Contribution to Skagit Prehistory. Unpublished Master's thesis, Department of Anthropology, Washington State University, Pullman.

Roll, T.E.

1974   *The Archaeology of Minard: A Case Study of a Late Prehistoric Northwest Coast Procurement System*. Unpublished Ph.D. dissertation, Department of Anthropology, Washington State University, Pullman.

Stewart, H.

1996   *Indian Artifacts of the Northwest Coast*. University of Washington Press. Seattle.

Wessen, G.C.

1989   *A Report of Archaeological Testing at the Dupont Southwest Site (45-PI-72), Pierce County, Washington*. Submitted to Weyerhaeuser Real Estate Company Land Management Division, Tacoma, Washington. Prepared by Western Heritage, Inc., Olympia, Washington.

Photo courtesy of Ray Pfortner.

9

Faunal Analysis: Mammal and Bird Remains

Kristine Bovy

As expected for a site located along the water's edge, most of the animal remains found at the Burton Acres Shell Midden are from animals living near the shoreline. Fish and shell-fish were the most abundant, but mammals such as mule deer, river otter and beaver were also found. Archaeologists make the assumptions that animal remains concentrated in a site reflect the diet of the people who lived there. Cooking is indicated by the fact that nearly ten percent of the mammals and bird bones were burned. Rodent bones, including mice, voles, and chipmunks, were also recovered. These mammals most likely burrowed into the site and were not part of a meal.

In contrast to the large number of identifiable fish bone described in the next chapter, there are considerably fewer mammal and bird remains recovered from the Burton Acres Shell Midden, most of which are highly fragmented and unidentifiable. Given that few archaeo-logical sites have been excavated in southern Puget Sound, the main objective of this analysis is to quantify the resources exploited by prehistoric inhabitants at this location. At least 50% of the bone is analyzed from each unit to obtain an accurate picture of non-fish faunal resources at the site. In addition, since the site was occupied before and after Euro-American contact, the bird and mammal assemblage was examined for significant change associated with that contact.

METHODOLOGY

Sampling

Units 22/58 and 26/57

Since the total number of bone fragments recovered from Units 22/58 and 26/57 is small, all of the bird and mammal material is analyzed from these units.

Units 28/58 and 29/58

The vertebrate faunal assemblage from Units 28/58 and 29/58 is larger than the other units, and therefore a 50% sample was taken to reduce analysis time. This analysis includes the material from the first bucket from every layer, plus all of the even-numbered buckets in that layer.

Sorting

Faunal remains with the same provenience (re-corded at the scale of 2- or 8-liter buckets, and one of four screen fractions) were sorted by taxonomic class; each class was put in a separate bag containing an identification label. The specimens were sorted into four categories: bird (Aves), mammal, indeterminate bird and/or mammal fragments, and indeterminate bone fragments (see Appendix F for details). More specific identifications are made for the mammal bone, if

---

Chapter opening photo: Archaeologist Mary Parr (left) shows this young volunteer how to map the location of his bucket. Volunteers first trowel the layer in which they are working and then collect the material using a dustpan. The carpenter rule measures from the edges of the unit to the precise spot collected.

possible, using comparative skeletal reference collection materials from the Zoology Division of the Burke Museum of Natural History and Culture, Seattle. In addition to taxonomic information, each specimen was inspected for evidence of characteristics such as burning and other modification. Descriptive and taphonomic attributes (reflecting the processes that act on organic remains after their death) were recorded (these are listed in Appendix F).

RESULTS

The results of the bird and mammal bone analysis are presented in three different ways. The descriptive summary lists those specimens which are identified to element. The quantitative section describes the abundance of bird, mammal and bird/mammal by unit and layers, and provides a brief comparison with the fish bone assemblage. Finally, the taphonomic summary includes information about burning and calcination, bone modification, fragmentation (screen-size data), and bird skeletal part representation. Further information on all the bone fragments analyzed is provided in Appendix F.

Descriptive Summary of Identifiable Specimens

This section provides descriptive information on all of the birds and mammalian specimens identifiable to skeletal element (e.g., indeterminate limb bone fragments are not listed). Where necessary, the criteria used to distinguish between closely related species are provided. The modern range distributions of the mammalian taxa represented in this assemblage are also discussed, primarily using information from Ingles (1965).

Class Aves (Birds)

Material: 14 skull fragments (2 basipterygoids, 1 frontal, 1 palatine, 3 pterygoids, 2 quadratojugals, 2 quadrates, 3 indeterminate skull fragments), 7 mandible fragments, 48 vertebrae (2 atlases, 13 cervicals, 13 thoracics, 5 centra, 15 indeterminate), 27 ribs (11 proximal, 1 proximal shaft, 1 distal, 2 sternal, 12 shafts), 2 innominates (acetabulum/ischium fragments), 7 synsacrum fragments, 7 coracoids, 3 furculae, 5 scapulae, 8 humeri (2 proximal, 5 distal, 1 shaft), 6 radii (3 proximal, 1 distal, 2 complete), 12 ulnae (2 proximal, 2 distal, 6 shaft, 2 complete), 1 carpal (cuneiform), 10 carpometacarpi (4 proximal, 2 distal, 4 complete), 2 first wing digits, 14 second wing digits (7

first phalanges, 7 second phalanges), 3 femora (1 proximal, 1 distal, 1 complete), 8 tibiotarsii (3 proximal, 2 distal, 1 distal epiphysis, 2 shaft), 2 tarsometatarsii (1 proximal shaft, 1 shaft), 14 phalanges (1 proximal, 8 distal, 5 complete).
Total: 200 specimens.
Remarks: The avian osteological terms used follow Howard (1929). In addition to the elements listed above, there are also 90 indeterminate bird limb bones, and 41 other indeterminate bird bone fragments. Table 9.6 (in taphonomic summary) shows the distribution of bird bones by unit. No attempt is made to identify the bird specimens to more specific taxonomic categories. Many appear to be duck-sized, however.

Class Mammalia (Mammals)

Table 9.1 lists the total number of mammalian specimens identified by unit, scientific, and common name. Since systematic refitting of bone fragments was not attempted, those fragments that fit together are counted as the actual number of fragments rather than as the number of refit specimens. The only exception to this approach is a shattered river otter skull found in Unit 26/57, which is counted as one specimen. Taxonomy and common names follow Wilson and Ruff (1999).

Mammalia, Unidentified

Material: 11 indeterminate skull fragments, 1 indeterminate tooth fragment, 7 vertebra fragments (2 centra, 1 epiphysis, 1 neural arch, 3 indeterminate), 33 ribs (shaft fragments), 1 humerus (proximal).
Total: 53 specimens.
Remarks: These specimens are determined to be mammal on the basis of their robusticity and/or morphological attributes, but they are not identifiable below the class level.

Order Rodentia (Rodents)
Rodentia, Unidentified

Material: 3 skull fragments (1 frontal, 1 premaxilla, 1 edentulous maxilla), 4 teeth (incisors), 3 ulnae (complete), 1 femur (proximal fragment).
Total: 11 specimens.
Remarks: Identification of incisors and postcranial rodent specimens was not attempted, and the skull fragments are too fragmentary or weathered to be identified to the family level.

Table 9.1 Identified Mammal Specimens (NISP)

Scientific Name	Common Name	Unit				Total
		22/58	26/57	28/58	29/58	
Rodentia, unidentified	rodent	1	1	8	1	11
Tamias townsendii	Townsend's chipmunk	-	-	-	1	1
Castor canadensis	beaver	-	-	-	2	2
Sigmodontinae, unidentified	New World rats and mice	1	-	2	-	3
Peromyscus sp.	deer mice	-	-	1	1	2
Arvicolinae, unidentified	mice, voles, muskrat	-	-	-	1	1
Clethrionomys gapperi	southern red-backed mouse	-	-	-	1	1
Microtus sp.	meadow voles	-	-	-	2	2
Carnivora, unidentified	carnivore	-	-	-	2	2
Procyon lotor	raccoon	-	-	1	1	2
Lontra canadensis	northern river otter	-	1	-	-	1
Odocoileus cf. *Hemionus*	mule deer	5	-	7	18	30
Unidentified Mammal	mammal	52	20	100	425	597
Total		59	22	119	455	655

Note: The northern river otter (*Lontra canadensis*) specimen consisted of a skull broken into at least 163 fragments.

Family Sciuridae (Squirrels)
Tamias townsendii (Townsend's chipmunk)

Material: 1 mandible (right mandible body with I, M_1 and M_2).

Total: 1 specimen.

Remarks: In addition to *Tamias townsendii* (Townsend's chipmunk), there are three other sciurids distributed on or near Vashon Island: the Western Gray squirrel (*Sciurus griseus*), Douglas's squirrel (*Tamiasciurus douglasii*), and the Northern Flying squirrel (*Glaucomys sabrinus*). These species are eliminated from consideration on the basis of morphology and their larger size. Townsend's chipmunks are currently distributed throughout western Washington and Oregon (Ingles 1965:186).

Family Castoridae (Beaver)
Castor canadensis (American Beaver)

Material: 2 teeth (lower incisor fragments).

Total: 2 specimens.

Remarks: These two lower incisor fragments are portions of a single tooth. Northwest Coast peoples hafted lower incisors of porcupine, muskrat, and especially beaver with antler to create a wood-carving tool. These teeth hold a sharp edge, and are resharpened easily (Stewart 1996:99). It is difficult to tell, however, whether wear on an incisor is due to use by humans for carving, or the actions of the beaver while living. Beavers were once very abundant in the Northwest Coast, but were nearly trapped out of existence in the nineteenth century to supply fur for hats and coats. Beavers are now re-established over most of their former range, and are quite numerous in certain locations along the coast (Ingles 1965:241).

Family Muridae (Rats, Mice, Voles)
Subfamily Sigmodontinae (New World Rats and Mice)
Sigmodontinae, Unidentified

Material: 2 skull fragments (1 left edentulous maxilla, 1 right edentulous maxilla), 1 mandible (left edentulous horizontal ramus).

Total: 3 specimens.

Remarks: These specimens are unidentifiable below the subfamily level. The members of this subfamily that inhabit the region today are: the deer mouse (*Peromyscus maniculatus*), the Northwestern deer mouse (*Peromyscus keeni* (=*oreas*)), and the bushy-tailed wood rat or packrat (*Neotoma cinerea*).

Peromyscus sp. (Deer Mice)

Material: 1 skull (right maxilla with M^1), 1 mandible (left mandible with I, M_1 and M_2).

Total: 2 specimens.

Remarks: There are two species of *Peromyscus* present

Table 9.2 Abundance (NISP) of Mammal, Bird, and Bird/Mammal Bone

Unit	Layer	Volume* (liters)	Mammal		Bird		Mammal/Bird		Total	
			C	C/V.	C	C/V	C	C/V	C	C/V
22/58	2A‡	36	1	0.0	-	-	-	-	1	0.0
	2B‡	142	1	0.0	-	-	-	-	1	0.0
	2C‡	60	7	0.1	1	0.0	6	0.1	14	0.2
	2D	58	8	0.1	3	0.1	1	0.0	12	0.2
	2E	48	27	0.6	-	-	8	0.2	35	0.7
	2F	22	2	0.1	-	-	2	0.1	4	0.2
	2G	16	12	0.8	-	-	3	0.2	15	0.9
	3A	24	1	0.0	-	-	-	-	1	0.0
	Unit Total	406	59	0.1	4	0.0	20	0.0	83	0.2
26/57	2A‡	102	3	0.0	-	-	1	0.0	4	0.0
	2B	36	181	5.0	1	0.0	28	0.8	210	5.8
	3A	8	-	-	-	-	-	-	-	-
	Unit Total	146	184	1.3	1	0.0	29	0.2	214	1.5
28/58	2A‡	68	49	0.7	54	0.8	66	1.0	169	2.5
	2B‡	76	52	0.7	59	0.8	98	1.3	209	2.8
	2C‡	64	12	0.2	12	0.2	11	0.2	35	0.5
	3A	24	6	0.3	1	0.0	2	0.1	9	0.4
	Unit Total	232	119	0.5	126	0.5	177	0.8	422	1.8
29/58	2A‡	78	61	0.8	25	0.3	29	0.4	115	1.5
	2B‡	42	29	0.7	34	0.8	32	0.8	95	2.3
	2C‡	84	92	1.1	110	1.3	187	2.2	389	4.6
	2D	56	100	1.8	28	0.5	47	0.8	175	3.1
	2E	70	157	2.2	9	0.1	21	0.3	187	2.7
	2F	48	16	0.3	4	0.1	2	0.0	22	0.5
	Unit Total	378	455	1.2	210	0.6	318	0.8	983	2.6
Total		1162	817	0.7	341	0.3	544	0.5	1702	1.5

Note: Indeterminate bone fragments (n=204), those not identifiable as bird, mammal, or fish, are not included.
‡ = layers thought to date to post-contact with Euro-Americans; all other layers date to pre-contact (see Table 5.1).
* = volume of excavated material that was analyzed for bird and mammal bone. C = count. C/V = count/volume.
Bird/Mammal = bone that is either bird or mammal, but cannot securely be distinguished.

in the Puget Sound region: the deer mouse (*Peromyscus maniculatus*) and the Northwestern deer mouse (*Peromyscus keeni*). These two species are so similar morphologically that they have only recently been shown to be two distinct species (see Wilson and Ruff 1999:571).

Subfamily Arvicolinae (Mice, Voles, Muskrat)
Arvicolinae, Unidentified
Material: 1 mandible (left edentulous horizontal ramus).
Total: 1 specimen.
Remarks: This edentulous specimen is not identifiable beyond the subfamily level. The arvicoline rodents present in the region are: the Southern Red-backed mouse (*Clethrionomys gapperi*), Western Heather vole (*Phenacomys intermedius*), Townsend's vole (*Microtus townsendii*), Long-tailed vole (*Microtus longicaudus*), Creeping vole (*Microtus oregoni*), and muskrat

(*Ondatra zibethicus*). Of these species, only the muskrat can be eliminated from consideration because of its larger size.

Clethrionomys gapperi (Southern Red-backed Mouse)
Material: 1 mandible (right body of mandible with M_1).
Total: 1 specimen.
Remarks: The Southern Red-backed mouse is currently distributed north of the Columbia River in grassy meadows, rock slides and chaparral zones (Ingles 1965:277). It is the only species of *Clethrionomys* present in Washington State.

Microtus sp. (Meadow Voles)
Material: 1 mandible (right mandible fragment with M_1), 1 tooth (right M_1).
Total: 2 specimens.
Remarks: *Microtus* species known from the vicinity of the Burton Acres Shell Midden site include the

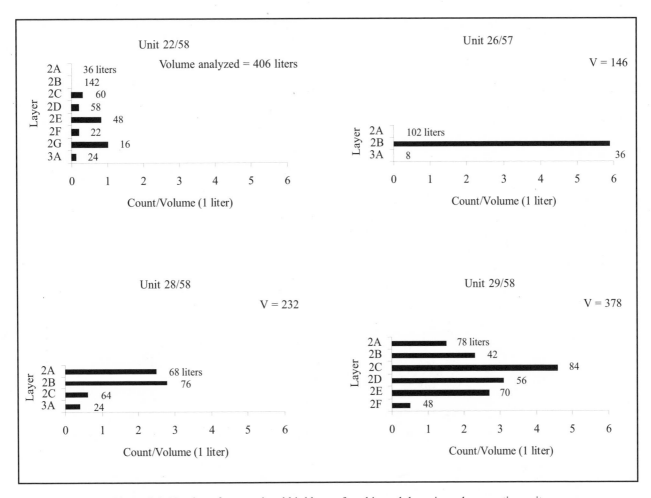

Figure 9.1 Number of mammal and bird bones found in each layer in each excavation unit.

Townsend's vole (*Microtus townsendii*), Long-tailed vole (*Microtus longicaudus*), and Creeping vole (*Microtus oregoni*). No attempt is made to distinguish between these morphologically similar species.

Order Carnivora (Carnivores)
Carnivora, Unidentified

Material: 2 teeth (1 crown fragment, 1 other indeterminate fragment).

Total: 2 specimens.

Remarks: These two indeterminate teeth fragments are coyote-sized or smaller. There are a number of medium-sized carnivores in the region, including coyotes (*Canis latrans*), domestic dogs (*Canis familiaris*), raccoons (*Procyon lotor*), bobcats (*Lynx rufus*) and a number of mustelids (e.g., marten, skunk, river otter).

Family Procyonidae (Raccoon)
Procyon lotor (Northern Raccoon)

Material: 1 tooth (premolar), 1 carpal (pisiform).

Total: 2 specimens.

Remarks: Raccoons are widely distributed in the Pacific region, usually living along lakes and other water margins (Ingles 1965:357). Native Northwest Coast peoples trapped raccoons both for their pelts and meat (Suttles 1951:96).

Family Mustelidae (Weasels, Otters)
Lontra canadensis (Northern River Otter)

Material: 1 crushed skull, including 6 teeth (1 canine, 2 premolars, 3 molars).

Total: 1 specimen.

Remarks: This crushed river otter skull was found in the SE 1/4 of the SE 1/4 of Unit 26/57, 16cm below the surface. The bone was noticed during excavation, and was taken out as a single field specimen. The skull is broken into at least 163 fragments. River otters occur throughout the Puget Sound area, occupying both

Table 9.3 Relative Abundance of Fish versus Mammal and Bird Bone

Unit	Layer	Fish		Mammal and Bird		Total
		C/V*	%	C/V.	%	C/V
22/58	2A‡	0.0	0.0	<0.1	100.0	<0.1
	2B‡	0.0	0.0	<0.1	100.0	<0.1
	2C‡	0.2	45.2	0.2	54.8	0.4
	2D	7.9	97.4	0.2	2.6	8.2
	2E	38.4	98.1	0.7	1.9	39.2
	2F	39.5	99.5	0.2	0.5	39.7
	2G	7.1	88.4	0.9	11.6	8.1
	3A	5.8	99.3	<0.1	0.7	5.7
	Unit Total	14.8	98.6	0.2	1.4	15.0
26/57	2A‡	<0.1	66.7	<0.1	33.3	0.1
	2B	1.4	19.4	5.8	80.6	7.2
	3A	1.3	100.0	0.0	0.0	1.3
	Unit Total	0.5	25.5	1.5	74.5	2.1
28/58	2A‡	5.4	68.5	2.5	31.5	7.9
	2B‡	24.0	89.7	2.8	10.3	26.8
	2C‡	11.9	95.6	0.6	4.4	12.5
	3A	3.2	89.4	0.4	10.6	3.6
	Unit Total	15.3	89.1	1.8	10.9	17.2
29/58	2A‡	6.9	82.4	1.5	17.6	8.4
	2B‡	12.4	84.5	2.3	15.5	14.6
	2C‡	8.4	64.4	4.6	35.6	13.0
	2D	15.9	83.6	3.1	16.4	19.1
	2E	52.3	95.1	2.7	4.9	55.0
	2F	88.1	99.5	0.5	0.5	88.5
	Unit Total	27.4	91.3	2.6	8.7	30.0
Total		19.1	92.9	1.5	7.1	20.6

‡ = layers thought to date to post-contact with Euro-Americans; all other layers date to pre-contact (see Table 5.1). C/V = count/volume (liters).
Note: The volume of excavated material analyzed varies between fish, and mammal and bird bone.

marine and freshwater habitats (Angell and Balcomb 1982:124). Suttles (1951:96) observed that Coast Salish peoples trapped river otters for their pelts using deadfall traps.

Order Artiodactyla (Even-toed Hoofed Mammals)
Odocoileus cf. *hemionus* (Mule Deer)

Material: 1 skull fragment (maxilla fragment with M³), 7 teeth (1 upper molar broken into 4 fragments, 1 other upper molar, 1 lower molar, 1 indeterminate fragment), 2 vertebrae (thoracic), 4 radii (1 right proximal, 2 left distal, 1 shaft), 1 ulna (proximal epiphysis), 2 carpals (right cuneiforms), 5 metacarpals (1 proximal broken into 5 fragments), 2 metapodials (shaft fragments), 2

tarsals (right cuneiform 2 & 3's), 4 phalanges (1 2nd, 1 3rd, 2 indeterminate).

Total: 30 specimens.

Remarks: These specimens cannot be distinguished from white-tailed deer (*Odocoileus virginianus*) on the basis of morphology. However, since mule (black-tailed) deer are currently distributed west of the Cascades, while white-tailed deer are not, they are most likely the former. Coast Salish people hunted black-tailed deer with bow and arrows, nets, and traps (Suttles 1951:82-91). Deer were important sources of meat, hide, marrow, and bone and antler for making tools (e.g., antler adze handles, ulna awls) (Suttles 1951:91).

Quantitative Summary

The abundance of mammal, bird, and indeterminate bird/mammal fragments from each excavation unit by layers is described in Table 9.2. The layers are divided into those deposited after contact with Euro-Americans and those deposited before contact with Euro-Americans. The volume corresponds to the number of liters of excavated material that was analyzed for bird and mammal bone. The counts of each different kind of bone are standardized by volume to obtain a measure of density that can be compared between units and layers. For example, in Unit 29/58, Layer 2C there is an average of 1.3 bird bone fragments per liter, while Unit 28/58 Layer 2C contain relatively less bird bone (0.2 bird bone fragments per liter). Almost half (48%, 817 specimens [n= 817]) of the identifiable bone is mammal. Next in abundance are indeterminate bird/mammal bone fragments (n=544), which comprise 32% of the identifiable bone at the site. Finally, there are 341 fragments of bird bone (20% of the assemblage). There are also 204 indeterminate bone fragments, those not identifiable as mammal, bird, or fish, which are not included in Table 9.2.

The abundance of all analyzed mammal and bird bone by unit is also shown graphically in Figure 9.1. These graphs show the total count of bird and mammal bone per liter for each layer. It is evident that the excavation units differ with respect to density of faunal material. Unit 22/58 has low levels (<.1) of mammal and bird fragments per liter. Unit 26/57, Layer 2B has a high density of bone fragments (5.0 count/liter) as a result of the crushed river otter skull. However, if all of the fragments (n = 163) that may fit with this skull are considered as one specimen, the mammal bone density

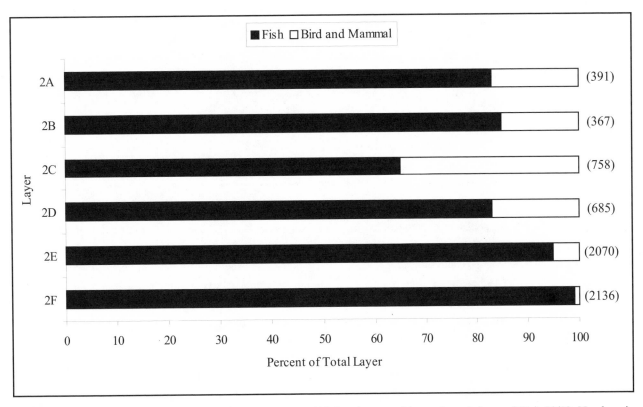

Figure 9.2 Number of fish bones compared to the number of bird and mammal bones in each layer of Unit 29/58. Numbers in parentheses represent total number of bone fragments per layer.

would be less than 1.0 fragments/liter. In Unit 28/58, Layer 2C has a relatively low number of mammal and bird bones per volume (.5 fragments/liter), with a marked increase in density in 2B (2.8 frag./liter). In contrast, in unit 29/58 the mammal and bird bone density peaks in Layer 2C (4.6 frag./liter), with a gradual decrease in the upper layers. In almost all cases, however, the density (ct./ vol.) of mammal and bird bone is considerably lower than that of fish bone (Table 9.3).

Figure 9.2 shows the relative frequencies of bird and mammal bone versus fish bone for Unit 29/58. Layers 2A, 2B, and 2C were deposited after contact with Euro-Americans, with the initial post-contact layer (2C) dating to the mid nineteenth to early twentieth century. Layers 2D/2E, and 2F were deposited prior to contact with Euro-Americans, beginning at about 900 years ago. Fish comprise over 80% of every layer except for Layer 2C, which is comprised of 65% fish and 35% bird and mammal bone.

A Chi-square test for independence is used to evaluate whether there is a significant difference in the distribution of classes of vertebrates through time. The null hypothesis is that the distribution of fish, bird, and

mammal bones is the same in the post-contact and pre-contact layers of Unit 29/58. The results are summarized in Table 9.4. Since a larger sample of excavated material was analyzed for bird and mammal bone, it is necessary to omit some of these data, so that the counts are comparable with the fish data. Specifically, the counts in Table 9.4 include the first and fourth bucket of each layer, plus every subsequent fourth bucket (e.g. bucket 1, 4, 8, 12...); the bird and mammal data obtained from the other even buckets (2, 6, 10...) are omitted.

The null hypothesis is rejected ($\chi^2 = 300.11$, p<.001): there is a highly significant difference in the distribution of fish, bird, and mammal bones in the post-contact and pre-contact layers. Analysis of the standardized residuals, a measure of how many standard deviations above or below the observed counts are from the expected counts, reveals that there are significantly more birds ($\varepsilon_j = 12.75$) and mammals ($\varepsilon_j = 8.16$) than expected in the post-contact layers.

Taphonomic Summary

Burning/Calcination

The presence of burning (identified by dark discoloration and/or surface cracking) and calcination (identified by

Table 9.4 Unit 29/58: Bone in Post-Contact and Pre-Contact Layers

Layers	Observed Count (Expected Count)			
	Fish	Bird	Mammal	Total
Post-Contact (2A, 2B, 2C)	917 (1036.20)	79 (20.8)	117 (56.0)	1113
Pre-Contact (2D, 2E, 2F)	4507 (4387.80)	30 (88.2)	176 (237.0)	4713
Total	5424	109	293	5826

$\chi^2 = 300.11$ (p<.001)

Table 9.5 Distribution of Burned and Calcined Bone Across Taxa

Taxon	Burned		Calcined	
	C	% of taxon	C	% of taxon
Mammal*	12	1	83	10
Bird	10	3	10	3
Bird/Mammal	24	4	23	4
Indeterminate	3	1	4	2
Total	49	3	120	6

C = count. * = includes one burned fragment of deer bone.

a whitened to bluish surface appearance) are recorded for all of the mammal, bird and indeterminate bones in the assemblage. Calcined bone is the result of heating above 600°C (Lyman 1994; McCutcheon 1992), and has a distinctive appearance that can rarely be confused with the results of other taphonomic processes. The black-to-brown discoloration of bone, however, can also be caused by the presence of certain chemical elements or compounds in the sediments in which they were buried (e.g., Shipman et al. 1984; McCutcheon 1992). Shipman et al. (1984) suggest that micro-morphological characteristics are more reliable than color for assessing the presence and degree of burning. Since calcination is present in the assemblage, it is likely that many of these blackened fragments are indeed burnt. However, it is possible that some of the specimens recorded as burnt may actually be chemically stained. Therefore, the counts of burned bone could be inflated.

Of the 1906 mammal, bird, and indeterminate bone fragments analyzed from the Burton Acres Shell Midden, 3% (n= 49) are recorded as burnt, and 6% (n=120) are identified as calcined. The percentages of burnt and calcined bone are between 2 and 5% for each of the individual excavation units, except for Unit 29/58 which has a slightly higher percentage of calcined fragments overall (9%). The frequency of calcination is greater in the post-contact layers of Unit 29/58 (13%, n=79), than in the pre-contact layers (3%, n=14). Table 9.5 shows the distribution of burned and calcined bone across taxa; a larger proportion (10%) of the mammal bone is calcined than the bird, bird/mammal or indeterminate bone fragments.

Other Cultural Modification

A few of the mammal and bird bones from the Burton Acres assemblage have evidence of other types of cultural modification. Specimens that are ground or have abrasion scratches are discussed in Chapter 8:

Bone and Antler Tools, while modification due to butchering is described here.

None of the bird or mammal bones analyzed from this assemblage have definite cut-marks made by stone tools. However, three mammal bones have been cut with a metal saw; one is a calcined fragment of a deer-sized (or smaller) limb bone shaft, broken into three fragments, with the saw marks on one edge running parallel to the shaft (Cat. # 29582C0062400), one is a fragment of a deer-sized bone sawed at both ends (Cat. # 29582A0152200), and the other is an unidentified cow-sized fragment of robust cancellous bone with two parallel saw-cut edges (Cat. # 28582B0182100). All of these saw-cut fragments were found in layers that also contain metal and glass artifacts.

In addition, four bone fragments (one deer-sized rib fragment, two deer-sized limb bone shaft fragments, and one indeterminate bird or mammal fragment) have flakes removed from the exterior surface of the bone. These flake scars may be the result of cultural modification or natural taphonomic processes.

Screen-Size Data / Fragmentation

During the excavation of the Burton Acres Shell Midden, archaeologists observed that the pre-contact layers contained more whole shells, while the shell in the post-contact layers were more fragmented (see Chapter 5). Phillips (Chapter 11) suggests this may be due to the presence of more fragile taxa in the upper layers, a change in harvesting or processing strategies, or post-depositional processes. Screen-size data for mammal and bird bone are investigated to determine if they also conform to this pattern.

Figure 9.3 shows the screen-size distribution of analyzed mammal and bird bone for Units 28/58 and 29/58 by layer. Data for Units 22/58 (n=83) and 26/57 (n=214) are not shown because the sample sizes in those

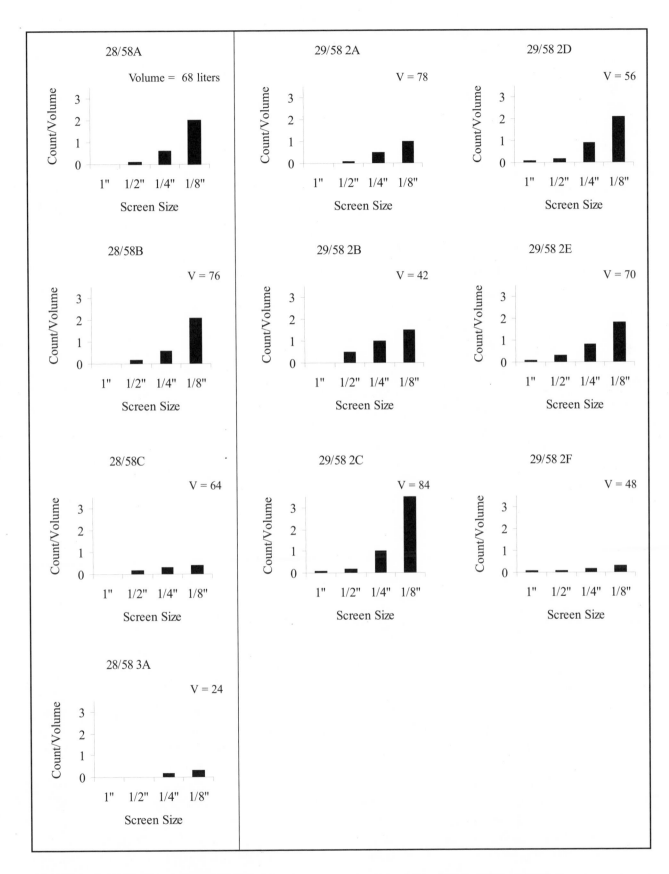

Figure 9.3 Number of mammal and bird bones found in each screen size for each layer in Units 28/58 and 29/58. Bars represent count per volume to standardize the comparison across screen sizes.

Table 9.6 Anatomical Part Distribution of Bird Bones

Skeletal Portion	Element	22/58	26/57	28/58	29/58	Total
Axial Skeleton	Skull	1	-	5	8	14
	Mandible	-	-	1	5	6
	Vertebra	-	-	13	35	48
	Rib	-	-	13	14	27
	Innominate	-	-	-	8	8
Total Axial		1	0	32	70	103
Appendicular Skeleton						
Pectoral Girdle	Sternum	-	-	-	-	0
	Scapula	-	-	-	5	5
	Coracoid	-	-	4	3	7
	Furculum	-	-	1	2	3
Total Pectoral Girdle		0	0	5	10	15
Wing	Humerus	-	-	3	5	8
	Radius	-	-	5	1	6
	Ulna	-	-	5	2	7
	Cuneiform	-	-	1	-	1
	Scapholunar	-	-	-	-	0
	Carpometacarpus	1	1	6	1	9
	Pollex	-	-	2	-	2
	Digit 2 Phalanx 1	-	-	4	3	7
	Digit 2 Phalanx 2	-	-	3	4	7
	Third Digit	-	-	-	-	0
Total Wing		1	1	29	16	47
Hind Limb	Femur	-	-	1	2	3
	Fibula	-	-	-	-	0
	Tibiotarsus	-	-	2	5	7
	Tarsometatarsus	-	-	-	2	2
	Phalanx	-	-	7	7	14
Total Hind Limb		0	0	10	16	26
Indeterminate Appendicular	Limb bone (indeterminate)	2	-	28	64	90
Total Appendicular		3	1	72	106	178
Indeterminate Fragments	Indeterminate	-	-	15	26	41
Total Bird Bone		4	1	119	202	322

Note: The numbers in this table take refit specimens into account, and therefore some of the numbers differ from the total NISP values listed in the descriptive summary.

units are comparatively small. The graphs for Unit 28/58 2C (n=35) and 3A (n=9), and Unit 29/58 2F (n=22) have a very flat curve because the sample sizes are small. It is evident from the graphs with larger sample sizes that all layers exhibit a similar pattern in which the count/volume (liters) increases markedly from 1 inch to 1/2 inch to 1/4 inch to 1/8 inch mesh sizes. This pattern is apparent throughout the archaeological sequence. Of the total mammal, bird, and indeterminate bone fragments recovered from the screens at the site, 0.5% (n= 9) are from 1 inch screens, 4% (n= 76) from 1/2 inch, 22% (n= 391) from 1/4 inch, and 73%

(n= 1267) from the 1/8 inch screen. While some of these 1/8 inch fragments are of rodents or small birds, most appear to be tiny fragments of larger-sized animal bones, indicating a high degree of fragmentation for the bird and mammal bone within the assemblage.

While the general screen size distribution pattern is similar throughout the cultural layers at the Burton Acres Shell Midden, differences do exist. For instance, there is a significant difference (α=.05) between the distribution of bones recovered from the 1 inch, 1/2 inch, 1/4 inch, and 1/8 inch screens in Unit 29/58 Layer 2C (post-contact) and Layer 2D/2E (pre-contact) (χ^2= 8.55, p=.036). In Layer 2C, the bones are more fragmented (there are higher than expected numbers recovered in the 1/8 inch screen).

Bird Skeletal Part Distribution

An unusual pattern of avian skeletal part distribution exists in numerous archaeological sites in the southern Northwest Coast, including the Beach Grove site in British Columbia (Matson et al. 1980), the Shoemaker Bay site on Vancouver Island (McMillan and St. Claire 1982), Tualdad Altu located along the Black River southeast of Seattle (Chatters 1988), and the West Point site in Seattle (Lyman 1995). In these sites, wing bones outnumber both leg and axial elements. For instance, at the Operation D portion of the British Camp site (45SJ24), San Juan Island, over half (n=112) of the total bird bones (n=197) identified to element from the prehistoric assemblage are elements of the wing (Bovy 1998).

The anatomical part distribution of birds from Burton Acres is shown in Table 9.6. In comparison to other sites in the southern Northwest Coast, this assemblage is unique in at least one respect. Axial elements (e.g., skull, vertebrae, ribs) are often rare in Northwest Coast avifaunal assemblages (e.g., Bovy 1998; Chatters 1988; Crockford et al. 1997; Ham et al. 1984; Seymour 1976). However, 54% of the identified elements in the Burton Acres assemblage are axial elements (n=103). One possible explanation for this difference is the recent age of these bird bones (most date to the post-contact occupation of the site). The bones may be better preserved or have different circumstances of deposition.

The avifaunal remains from Burton Acres are similar to other sites in the region, however, in terms of the relative proportion of wing and leg elements. The Burton Acres avifaunal assemblage contains 47 wings and 26 legs. In a whole bird, there are usually 20 bones of the wing: humerus (2), radius (2), ulna(2), scapholunar (2), cuneiform (2), carpometacarpus (2), digit 2 phalanx 1 (2), digit 2 phalanx 2 (2), pollex (2), and third digit (2), and 36 leg bones: femur (2), tibiotarsus (2), fibula (2) tarsometatarsus (2), and 48 phalanges. The expected wing-to-leg ratio for a whole bird is therefore 20/36 or 0.6. A χ^2 "goodness of fit" test reveals that the wing and leg counts at the Burton Acres site do not conform to this expected proportion ($\chi^2 = 26.13$, p<.001); there are more wings than expected.

Researchers have suggested many hypotheses for the abundance of wings in archaeological assemblages, including curation of wings for tool use (e.g., DePuydt 1994), processing and consumption practices (e.g., Hanson 1991), differential transport (e.g., Ham 1982), and differential preservation (e.g., Hanson 1991). Elsewhere (Bovy 2002), I have summarized these hypotheses and conducted a preliminary test of the differential preservation hypothesis, which suggests that wing bones survive better than leg bones because they are denser. I conclude that this density hypothesis is not well supported by the available data.

DISCUSSION

The goal of this analysis is to provide a quantitative summary of the mammal and bird resources at the Burton Acres Shell Midden. In addition, the subsistence practices at the site were examined to identify alterations resulting from Euro-American contact (e.g., was there a change in the species taken?). This second goal is difficult to evaluate because of the low number of identifiable fragments in the assemblage. In addition, mammal and bird resources appear to play a minor role in the subsistence of the site throughout its occupation.

The bird and mammal bones from this site are quite fragmented, making identification difficult. The majority (73%) of the fragments recovered from the screens are from the 1/8 inch, and most of these are broken fragments of medium- to large-sized mammals. While the screen-size data (Figure 9.3) indicates a similar overall pattern of fragmentation through time, there is a significant increase in the 1/8 inch material recovered in Layer 2C (post-contact) of Unit 29/58, relative to the previous pre-contact layer (2D/2E). This corresponds to the increased fragmentation of shell in the post-contact layers. Whether this fragmentation occurred before or after deposition is not clear (see discussion in Chapter 13).

Only 58 of the mammal specimens (9% of the total) are identifiable below the class level, and of these, 23 are rodents. Mule deer, the taxa which has the highest number of identifiable specimens (n=30), are present throughout the layers, as are rodents. Although identification of the bird bone is not attempted below the class level, the bird assemblage appears to be well preserved compared to the mammal assemblage. Therefore, future identification may provide more specific data about temporal changes in the frequencies of bird species.

The bird and mammal assemblage from the Burton Acres Shell Midden comprises a small proportion of the faunal resources throughout the archaeological sequence; the density of bird and mammal bone at the site is relatively low in comparison to the density of fish bone (Table 9.3) and shellfish (Table 11.1). An analysis of the distribution of fish, bird, and mammal bone in Unit 29/58 indicates that there are significant differences in the proportions of these vertebrates in the pre- and post-contact layers (χ^2= 300.118, p<.001). There is a significant increase in the number of birds and mammals in the post-contact layers, while the number of fish decrease. This increase may reflect a real change in the relative importance of these taxa, or may simply reflect the increased fragmentation of the mammal and bird assemblages in the post-contact layers. Kopperl and Butler (Chapter 10) observed a decrease in the importance of herring, the primary fish taxon exploited at the site, through time.

Aside from the significantly modified bone objects described in Chapter 8, the only clear indications of human modification of the bird and mammal assemblage are three metal saw-cut mammal bone fragments. None of the specimens analyzed exhibited cut marks made by stone tools. The lack of stone tool cutmarks is not surprising given the small sample size of the assemblage (e.g., Lyman 1987) and the high degree of fragmentation.

Burned and calcined bone (approximately 9% of the bird and mammal assemblage) is likely the result of human modification, rather than natural burning, given the context of the bone within the shell midden. The percentage of calcined bone increased from 3% in the pre-contact layers of Unit 29/58 to 13% in the post-contact layers. None of the rodent bones within the Burton Acres assemblage are modified by fire or other cultural modifications. These rodents probably reflect either post-depositional disturbance (e.g., burrowing rodents such as the Townsend's chipmunk), or natural deposition at the time of past human occupation. Only six bird bones from Burton Acres Shell Midden appear to be either burnt or calcined, leaving open the possibility of natural deposition of bird bone. Bird bone is deposited naturally by being washed up on shore or through predation by other animals.

CONCLUSION

There are 1906 mammal, bird and indeterminate bone fragments from the Burton Acres Shell Midden, which are analyzed in terms of taxonomic affiliation, abundance, and modification. All of the mammal and bird bone from Units 22/58 and 26/57 are analyzed, while a 50% sample was taken from Units 28/58 and 29/58. The bone is sorted into four main categories: mammal (n= 817), bird (n= 341), bird/mammal (n= 544), and indeterminate (n= 204), and the mammal bone was identified as specifically as possible with the aid of comparative collections. Only a small percentage (9%) of the mammal bone is identifiable. The following mammalian taxa are identified (in order of abundance): mule deer, unidentified rodent, unidentified sigmodontid (New World rats and mice), American beaver, deer mouse, meadow vole, unidentified carnivore, Northern raccoon, Northern river otter, Townsend's chipmunk, arvicoline rodent, and Southern red-backed mouse.

The density of bird and mammal bone is low throughout the archaeological record (average of 1.5 bone fragments per liter), with mammal bone the most abundant (0.7 frag./liter), followed by ambiguous bird/mammal fragments (0.5 frag./liter) and bird bone (0.3 frag./liter). With the exception of Unit 26/57, which contains the highly fragmented river otter skull, the density of the total mammal and bird bone is much lower than fish bone. Of the total mammal, bird and indeterminate bone analyzed, 3% is blackened (possibly burnt), and 6% is calcined. There are no cut marks on the bone fragments, although three mammal specimens were cut with a metal saw. Aside from burning, there is little evidence for human modification, and indeed the rodents (and possibly some of the bird specimens) may reflect natural deposition. The assemblage is highly fragmented throughout the archaeological sequence, with 73% of the screened material recovered from 1/8 inch screens.

Throughout the occupation of the Burton Acres Shell Midden location, the mammal and bird bone accumulation has been relatively consistent. The only quantifiable change is an increase in the amount of burning and fragmentation in the post-contact layers. A similar change is found in the shell remains (Chapter 11). Overall mammals and birds appear to have been much less important than fish to the pre- and post-Euro-American contact occupants of this site.

REFERENCES

Angell, T., and K.C. Balcomb III

1982 *Marine Birds and Mammals of Puget Sound.* Puget Sound Books. University of Washington Press, Seattle.

Bovy, K.M.

1998 Avian Skeletal Part Distribution in the Northwest Coast: Evidence from the British Camp Site, Op-D (45-SJ-24). Unpublished Master's thesis, Department of Anthropology, University of Washington, Seattle.

2002 Differential Avian Skeletal Part Distribution: Explaining the Abundance of Wings. *Journal of Archaeological Science*, in press.

Chatters, J.C.

1988 *Tualdad Altu (45KI59): A 4th Century Village on the Black River, King County, Washington.* First City Equities, Seattle.

Crockford, S., G. Frederick, and R. Wigen

1997 A Humerus Story: Albatross Element Distribution from Two Northwest Coast Sites, North America. *International Journal of Osteoarchaeology* 7:287-291.

DePuydt, R.T.

1994 Cultural Implications of Avifaunal Remains Recovered from the Ozette Site. In *Ozette Archaeological Project Research Reports, Vol. II: Fauna*, edited by S.R. Samuels, pp. 197-263. Reports of Investigations No. 66. Department of Anthropology, Washington State University, Pullman and National Park Service, Pacific Northwest Regional Office.

Ham, L.C.

1982 Seasonality, Shell Midden Layers, and Coast Salish Subsistence Activities at the Crescent Beach Site, DgRr 1. Unpublished Ph.D. dissertation, Department of Anthropology and Sociology, University of British Columbia, Vancouver.

Ham, L.C., A.J. Yip, and L.V. Kullar

1984 *The 1982/83 Archaeological Excavations at the St. Mungo Site (DgRr 2), North Delta, British Columbia*, vol. 1. Consultant's report on file, Ministry Library, Ministry of Small Business, Tourism and Culture, Victoria, British Columbia.

Hanson, D.K.

1991 Late Prehistoric Subsistence in the Strait of Georgia Region of the Northwest Coast. Unpublished Ph.D. dissertation, Simon Fraser University, Department of Archaeology, Burnaby, British Columbia.

Howard, H.

1929 *The Avifauna of Emeryville Shellmound.* California University Publications in Zoology Vol. 32. University of California Press, Berkeley.

Ingles, L.

1965 *Mammals of the Pacific States: California, Oregon, Washington.* Stanford University Press, Stanford, California.

Lyman, R.L.

1987 Archaeofaunas and Butchering Studies: A Taphonomic Perspective. In *Advances in Archaeological Method and Theory* vol. 10, edited by M. Schiffer, pp. 249-337. Academic Press, New York.

1994 *Vertebrate Taphonomy.* Cambridge University Press, New York.

1995 Mammalian and Avian Zooarchaeology of the West Point, Washington Archaeological Sites (45KI428 and 45KI429). In *The Archaeology of West Point, Washington: 4,000 Years of Hunter-Fisher-Gatherer Land Use in Southern Puget Sound*, edited by L.L. Larson and D.E. Lewarch. Submitted to CH2M Hill, Bellevue, Washington. Prepared for King County Department of Metropolitan Services. Prepared by Larson Anthropological/Archaeological Services.

Matson, R.G., D. Ludowicz, and W. Boyd

1980 *Excavations at Beach Grove in 1980.* Report to Heritage Conservation Branch, Victoria, British Columbia. Museum of Anthropology, and Anthropology and Sociology Department, University of British Columbia, Vancouver.

McCutcheon, P.T.

1992 Burned Archaeological Bone. In *Deciphering a Shell Midden*, edited by J.K. Stein, pp. 347-370. Academic Press, San Diego.

McMillan, A.G., and D.E. St. Claire

1982 *Alberni Prehistory: Archaeological and Ethnographical Investigations on Western Vancouver Island.* Theytus Books, Penticton, British Columbia.

Seymour, B.

1976 1972 Salvage Excavations at DfRs3, the Whalen Farm Site. In *Current Research Reports*, edited by R. Carlson, pp. 83-98. Publication No. 3. Department of Archaeology, Simon Fraser University, Burnaby, British Columbia.

Shipman, P., G. Foster, and M. Schoeninger

    1984   Burnt Bones and Teeth: an Experimental Study of Color, Morphology, Crystal Structure and Shrinkage. *Journal of Archaeological Science* 11:307-325.

Stewart, H.

    1996   *Stone, Bone, Antler and Shell: Artifacts of the Northwest Coast.* University of Washington Press, Seattle.

Suttles, W.P.

    1951   Economic Life of the Coast Salish of Haro and Rosario Straits. Unpublished Ph.D. dissertation, Department of Anthropology, University of Washington, Seattle.

Wilson, D.E., and S. Ruff (editors)

    1999   *The Smithsonian Book of North American Mammals.* Smithsonian Institution, Washington, D.C.

10

Faunal Analysis: Fish Remains

Robert Kopperl and
Virginia Butler

Anyone who has been to Burton Acres Park will notice the constant wind that makes it an ideal place to dry fish. Not surprisingly, fish remains were found in abundance in this shell midden. Salmon were, and are, one of the most important foods for the people of the Puget Sound. At the Burton Acres Shell Midden, however, salmon takes a notable second place to herring. This small fish comprises nearly 80% of the fish found in the layers. Burned fish bone is further evidence of drying and cooking. Many remains found in the archaeological deposits are from fish that no longer live in Quartermaster Harbor, victims of destroyed habitat and pollution.

Excavation, recovery and analysis of almost 9000 fish remains from the Burton Acres Shell Midden provides an opportunity to address the question of prehistoric fish use in southern Puget Sound. While the fish fauna is represented by a wide variety of fish common to the Puget Sound today, herring (Clupeidae) dominate the assemblage. The abundance of fish bone at this site, especially herring, reflects a significant reliance on aquatic resources besides shellfish. From this analysis, the spatial distribution of different taxa and the condition of the remains that have been recovered provide a picture of changing fish utilization at the site, and may indicate the changing subsistence patterns of Native peoples as a result of European contact.

METHODS

Sorting and Sampling

Given time constraints, all of the fish remains could not be analyzed. The sampling strategy was designed to provide even spatial coverage and allow our results to be comparable to other project analyses. The fish remains examined were recovered from approximately 25% of the buckets excavated, selecting every first and fourth bucket from each excavation level. Specimens were recovered from all four mesh sizes (1 inch, 1/2 inch, 1/4 inch, and 1/8 inch) used during excavation.

Analysis Protocol

Fish remains were identified to the finest taxonomic level possible using the comparative collections of Virginia Butler (Portland State University), Mike Etnier (University of Washington), and the National Marine Mammal Laboratory at the Sand Point National Oceanic and Atmospheric Administration (NOAA) station in Seattle, Washington. Besides taxonomic information, specimens were visually examined for evidence of burning. A coding system was developed (see Appendix

Chapter opening photo: Archaeologist Bob Kopperl (center) gives instructions to two volunteers, while friends watch over their shoulders. Fish bones were found in such large numbers at the Burton Acres Shell Midden that people had to be reminded to keep excavating and not stop to identify individual bones while excavating. It was much more efficient to identify the hundreds of fish bones after the bucket was screened and contents spread on sorting trays.

G) for entering data from the fishbone analysis onto a database. Certain pieces of information were entered for each specimen examined, including a code for the finest taxon to which it could be identified, a code for the element (bone), whether or not it has a diagnostic landmark, whether or not the specimen is burned, and the quantity of specimens that share this same information from a particular provenience (for example, the number of burned herring caudal vertebrae from the same bucket and mesh size).

Analysis commenced at the Burke Museum of the University of Washington. Each bag of fish bones, which had an individual catalogue number, was separated into elements, burned and unburned, from different taxa. Each group (a particular element from a particular taxa, either burned or unburned) is a separate entry in the database. All fish remains, bagged by excavation unit, level, bucket and screen fraction, were rebagged by taxon. Blank forms with spaces for all the variables were created on Microsoft Excel, filled in by hand, entered into an SPSS (Statistical Package for the Social Sciences) file using the same variables, and saved both in SPSS for statistical analysis and Excel for compact data tables (see Appendix G).

Fish remains are quantified using NISP, or Number of Identified Specimens, which is the quantity of all remains identified for each taxon. Each specimen, whether broken or unbroken, is treated as a single NISP. Other methods of quantification were not used, such as MNI, or Minimum Number of Individuals. This counting unit is the minimum number of individuals represented by a taxon (quantified by either the left or the right most abundant elements, or just the number if the element is unpaired, such as vertebrae). Grayson (1984) and Butler (1987) have shown that MNI and NISP are highly correlated. Given the aggregation problems associated with MNI, NISP is used to quantify taxonomic abundance in this analysis.

DESCRIPTIVE SUMMARY

This section provides a description of the fish taxa identified in the Burton Acres Shell Midden. The data discussed in this section are displayed in Table 10.1. The first column provides the scientific name of each taxon found at the site. Each of these taxa are described in further detail below. The second column gives the common name of each taxon. The last five columns

provide the quantity of each taxon found in each excavation unit, as well as the total number of fish bones from each taxon found at the site.

The descriptive summary below details each fish taxon found at 45KI437. Of the almost 9000 fish bones analyzed, 5321 specimens are identified to at least the taxonomic level of order. The scientific name for each taxon, including class, order and family, genus, and species where appropriate, is given. A list of the skeletal elements found for each taxon is followed by remarks regarding how these elements are identified, and pertinent ecological and ethnographic information where available. Also included are data regarding each taxon's particular distribution in modern-day Puget Sound. This information comes from a report published by Miller and Borton (1980) for the Washington Sea Grant, who compiled a series of distribution maps for every species of fish reported in Puget Sound, based on reports from both private citizens and government agencies.

Class Chondrichthyes (cartilaginous fish)
Subclass Elasmobranchii (sharks, rays)
Order Squaliformes
Family Squalidae
Squalus acanthias (Spiny Dogfish)
Material: 15 vertebrae fragments.
Total: 15 specimens.
Remarks: Spiny dogfish vertebrae have a very distinctive spool-shape, which allows for species-level identification when encountered. They are the most abundant of seven shark Families and ten shark species in Puget Sound (Miller and Borton 1980). They inhabit both shallow and deep water and feed on small fish such as herring and smelt (Hart 1973).

Order Rajiformes (skates and rays)
Family Rajidae
Raja sp. (Skate)
Materials: 29 teeth, 1 dermal denticle.
Total: 30 specimens.
Remarks: Skate teeth have distinctive mushroom-shaped bases and their dermal denticles have flared bases with serrated edges. There are four species of skate in Puget Sound, all being shallow bottom-feeders. Comparative material was limited, so identification of these specimens is made at the genus-level. *Raja*

Table 10.1 Identified Fish Specimens (NISP)

Scientific Name	Common Name	Unit				Total
		22/58	26/57	28/58	29/58	
Squalus acanthias	Spiny dogfish	3	-	4	8	15
Raja sp.	Skate	7	-	1	22	30
Hydrolagus colliei	Spotted ratfish	4	-	4	9	17
Clupea harengus pallasi	Pacific herring	427	4	1011	2838	4280
Oncorhynchus sp.	Salmon	9	-	291	268	568
Porichthys notatus	Plainfin midshipman	-	-	1	5	6
Family Gadidae	Codfish	-	-	2	6	8
Family Embiotocidae	Surfperch	24	-	2	55	81
Embiotoca lateralis	Striped seaperch	2	-	-	6	8
Rhacochilus vacca	Pile perch	-	-	-	3	3
Family Scorpaenidae	Rockfish	4	-	-	24	28
Family Cottidae	Sculpin	5	-	4	34	43
Leptocottus armatus, cf.	Pacific staghorn sculpin	-	-	-	14	14
Scorpaenichthys marmoratus, cf.	Cabezon	-	-	-	2	2
Order Pleuronectiformes	Flatfish	69	1	21	116	207
Family Pleuronectidae	Right-eye flatish	-	-	-	2	2
Lepidopsetta bilineata, cf.	Rock sole	1	-	-	1	2
Microstomus pacificus, cf.	Dover sole	-	-	-	1	1
Parophrys vetulus, cf.	English sole	-	-	-	3	3
Platichthys stellatus, cf.	Starry flounder	-	-	-	3	3
Total		555	5	1341	3420	5321

binoculatas (big skate) and *Raja rhina* (longnose skate) are by far the most common skates in the area and have been recorded in Quartermaster Harbor (Miller and Borton 1980).

Order Chimaeriformes
Family Chimaeridae (Chimeras)
Hydrolagus colliei (Spotted ratfish)

Materials: 17 teeth.

Total: 17 specimens.

Remarks: The teeth of the spotted ratfish are distinctive wavy bony plates with vertical ridges perpendicular to the cutting edge. The spotted ratfish is the only member of the order Chimaeriformes found along the Northwest Coast (Hart 1973). Therefore, identification is made at the species level. The fish often inhabits deep waters but has been known to exhibit seasonal variation in its depth in Puget Sound. This occurs mainly in the spring when the spotted ratfish shows greater abundance in shallower water (Quinn et al. 1980).

Class Osteichthyes (bony fishes)
Order Clupeiformes
Family Clupeidae
Clupea harengus pallasi (Pacific herring)

Materials: 42 articulars, 1 basihyal, 24 basioccipitals, 4 basypterygia, 43 ceratohyals, 23 cleithra, 63 dentaries, 5 ectopterygoid, 27 epihyals, 43 exoccipitals, 37 frontals, 32 hyomandibulae, 25 upper hypohyals, 2 lower hypohyals, 37 interopercles, 40 maxillae, 1 metapterygoid, 68 opercles, 12 parasphenoids, 15 parietals, 30 postcleithra, 19 posttemporals, 63 prefrontals, 39 preopercles, 345 prootics, 67 pterotics, 37 quadrates, 10 scapulae, 42 subopercles, 34 supraoccipitals, 5 urohyals, 7 vomers, 16 1st vertebrae, 916 abdominal vertebrae, 1817 caudal vertebrae, 35 hypurals, 11 miscellaneous spines, 8 unidentifiable fragments.

Total: 4280 specimens.

Remarks: Herring is by far the most abundant fish taxon in the Burton Acres Shell Midden assemblage.

There are two species of clupeids found in the northeast Pacific: *Clupea harangus pallasi* (Pacific herring) and *Sardinops sagax* (Pacific sardine) (Hart 1973). The Pacific sardine is very rare in Puget Sound, whereas a herring run has been reported in the immediate vicinity of Burton Acres Shell Midden in Quartermaster Harbor (Miller and Borton 1980). Pacific herring are an anadromous schooling fish that migrate to shallow water in the late winter to spawn into the early spring (Hart 1973). In all likelihood, the archaeological material from 45KI437 represents Pacific herring, but given that comparative material from *Sardinops sagax* was not available for review, the identification must remain somewhat tentative.

Order Salmoniformes
Family Salmonidae
cf. *Oncorhynchus* sp. (Salmon)

Materials: 1 basioccipital, 1 ceratohyal, 1 cleithrum, 2 dentaries, 1 exoccipital, 1 parasphenoid, 1 parietal, 7 radials, 1 scapula, 1 suborbital, 2 supracleithra, 17 teeth, 3 type 1 (1st) vertebrae, 30 type 2 (abdominal) vertebrae, 57 type 3 (caudal) vertebrae, 2 type 4 (ultimate) vertebrae, 425 indeterminate vertebrae, 2 hypurals, 12 unidentifiable fragments.

Total: 568 specimens.

Remarks: Salmonid vertebrae and many other skeletal elements are easy to identify due to their morphology (Cannon 1987). Abdominal and caudal vertebrae can be distinguished on the basis of caudal and haemel processes (Butler 1993). However, discrimination beyond genus level is very difficult to accomplish without otoliths, as in the case for this assemblage. Butler (1987) developed a discriminant function analysis that uses measurements on 1st vertebrae of salmon to statistically assign a specimen to a particular species. With only three first vertebrae from this site, all in fragmentary condition, identification below genus-level is not possible.

The waters of Puget Sound and surrounding coastal streams are home to seven species in the genus *Oncorhynchus* (salmon and trout), two species of *Salvelinus* (*S. malma* - dolly varden; *S. confluentus* - bull trout), and one species of whitefish (*Prosopium williamsoni* - mountain whitefish) (Miller and Borton 1980). The remains from Burton Acres Shell Midden could be from any of these species, although in general,

their large size suggests that the archaeological sample is from *Oncorhynchus* (salmon and trout). However, chinook salmon (*Oncorhynchus tshawytscha*) are the only reported salmon to be observed in the vicinity of Quartermaster Harbor (Miller and Borton 1980).

Salmon are anadromous fish that spend part of their lives growing to maturity in fresh waters, part of their lives in marine waters, then returning to the same fresh water system to spawn. Salmon are common throughout the Pacific Northwest in most marine environments, and have been extensively documented as being utilized ethnographically and historically, and are found in most archaeological sites.

Order Batrachoidiformes
Family Batrachoididae (toadfish)
Porichthys notatus (Plainfin midshipman)

Materials: 1 abdominal vertebra, 2 caudal vertebrae, 3 indeterminate vertebrae.

Total: 6 specimens.

Remarks: The vertebrae identified to this taxa display neural and haemal processes that are diagnostic of Batrachoididae. Identification beyond family-level to species is made because no other species of Batrachoididae has been found north of Point Conception, California (Butler 1987). In Puget Sound, the plainfin midshipman is widespread, dwelling in both shallow intertidal waters where spawning occurs in the spring and in deeper waters at other times of the year (Miller and Borton 1980; Hart 1973).

Order Gadiformes
Family Gadidae (Codfish)

Materials: 1 otolith, 4 abdominal vertebrae, 1 caudal vertebra, 2 indeterminate vertebra.

Total: 8 specimens.

Remarks: Although there are numerous otoliths found in the Burton Acres Shell Midden assemblage, only one could be identified to family Gadidiae. The others are too fragmentary to determine taxon. The vertebrae are identifiable to family as well. Despite the availability of all four species of codfish found in Puget Sound available in the comparative collections, any finer identification is difficult because the bones are too fragmented. The four species of codfish are Pacific cod (*Gadus macrocephalus*), Pacific hake (*Merluccius productus*), Pacific tomcod (*Microgadus proximus*), and walleye pollock (*Theragra chalcogramma*). All four have been found in Quartermaster Harbor (Miller and Borton 1980).

Order Scorpaeniformes
Family Scorpaenidae (Scorpionfish and Rockfish)

<u>Materials</u>: 1 dentary, 1 exoccipital, 1 hyomandibular, 3 abdominal vertebrae, 15 caudal vertebrae, 7 indeterminate vertebrae.

<u>Total</u>: 28 specimens.

<u>Remarks</u>: Because of the limited comparative material available, identification of rockfish elements is to family. There are two genera and 27 species of rockfish found in Puget Sound today (Miller and Borton 1980). The species most commonly found near Vashon Island today are brown rockfish (*Sebastes auriculatus*), copper rockfish (*Sebastes caurinus*), yellowtail rockfish (*Sebastes flavidus*), quillback rockfish (*Sebastes maliger*), black rockfish (*Sebastes melanops*), bocaccio (*Sebastes paucispinis*), and the canary rockfish (*Sebastes pinniger*) (Miller and Borton 1980). Rockfish are bottom fish that inhabit waters ranging from intertidal zones to very deep open water (Hart 1973).

Family Cottidae (Sculpin)

<u>Materials</u>: 1 articular, 1 upper hypohyal, 2 opercles, 1 palatine, 2 premaxillae, 1 quadrate, 2 radials, 3 abdominal vertebrae, 29 caudal vertebrae, 1 hypural.

<u>Total</u>: 40 specimens.

<u>Remarks</u>: There are 36 species of sculpin known to inhabit Puget Sound today (Miller and Borton 1980). The genera of Cottidae available in the comparative collections used are *Leptocottus* sp., *Cottus* sp., *Myoxocephalus* sp., *Enophrys* sp., *Scorpaenichthys* sp., *Hemilepitotus* sp., and *Chitonotus* sp. Unfortunately, only a few elements such as some of the jaw parts and some cranial elements, especially the preopercle, can be identified to species-level, and these assignments are based on exact matches with the limited comparative material available. This makes those identifications rather tentative. The rest of the elements, mainly vertebrae and fragmented cranial elements, have only been identified to family-level.

cf. *Leptocottus armatus* (Pacific staghorn sculpin)

<u>Materials</u>: 2 basioccipitals, 1 frontal.

<u>Total</u>: 3 specimens.

<u>Remarks</u>: These identifications are tentative because of the number of Cottidae genera that exist and the limited comparative material. The Pacific staghorn sculpin is perhaps the most widespread sculpin in Puget Sound. Besides this species, there is only one other sculpin reported in Quartermaster Harbor: the slim sculpin (*Radulinus asprellus*) (Miller and Borton 1980). The slim sculpin has a very gracile skeleton (hence the name), and is on average a third of the size of the Pacific staghorn sculpin (Hart 1973), which aided in the identification of these elements as belonging to the genus *Leptocottus*.

Leptocottus armatus (Pacific staghorn sculpin)

<u>Materials</u>: 1 dentary, 5 premaxillae, 1 preopercle, 3 vomers, 1 miscellaneous jaw fragment.

<u>Total</u>: 11 specimens.

<u>Remarks</u>: The elements of the mouth region, and especially the antler-shaped preopercle of the Pacific staghorn sculpin, are diagnostic (Wydoski and Whitney 1979; Butler 1987). Based on their morphology, these elements from the Burton Acres Shell Midden assemblage have been assigned to *Leptocottus armatus*.

cf. *Scorpaenichthys marmoratus* (Cabezon)

<u>Materials</u>: 1 dentary, 1 frontal.

<u>Total</u>: 2 specimens.

<u>Remarks</u>: Identification of the cranial fragments is from comparative collections at the National Marine Mammal Lab at NOAA, Seattle, WA. The specimens from Burton Acres Shell Midden are from an extremely large sculpin. There are only two species of sculpin that attain this size: the great sculpin (*Myoxocephalus polyacanthocephalus*) and cabezon (*Scorpaenichthys marmoratus*) (Hart 1973). The archaeological specimens were morphologically identical to the cabezon from the comparative material, most notably in proportional tooth-socket width and in the angle between the toothed portion and the inferior portion of the dentary. The size and patterning on the frontal bone is identical to the cabezon as well. The cabezon is one of the largest sculpins, and is probably the largest to inhabit Puget Sound. It has been reported in the area around Vashon Island but not within Quartermaster Harbor proper (Miller and Borton 1980). They inhabit water of moderate depths although they are also found in shallow water (Hart 1973).

Order Perciformes
Family Embiotocidae (Surfperches)

<u>Materials</u>: 1 quadrate, 1 vomer, 5 lower pharyngeals, 58 pharyngeal teeth, 1 symplectic, 2 miscellaneous jaw fragments, 6 abdominal vertebrae, 5 caudal vertebrae, 1 indeterminate vertebra, 1 unidentifiable fragment.

<u>Total</u>: 81 specimens.

<u>Remarks</u>: Identification of surfperches to species can only be done with confidence to mouthparts, especially pharyngeals and premaxillae, as well as the parasphenoid. Teeth from surfperch pharyngeals are fairly common in the assemblage but cannot be assigned to species in this analysis based on the available comparative material. This was the case with the elements in this category. There are six species of surfperch that inhabit Puget Sound today, only three of which are common near Quartermaster Harbor: shiner perch (*Cymatogaster aggregata*), striped seaperch (*Embiotoca lateralis*), and pile perch (*Rhacochilus vacca*) (Miller and Borton 1980). Bays and intertidal zones, especially near rocky shores and old piers, are common habitat for surfperch (Hart 1973).

Embiotoca lateralis (Striped seaperch)

<u>Materials</u>: 2 premaxillae, 4 lower pharyngeals, 2 upper pharyngeals.

<u>Total</u>: 8 specimens.

<u>Remarks</u>: There are no modern reports of striped seaperch in Quartermaster Harbor, but they have been found nearby in Colvos and Dalco Passages (Miller and Borton 1980).

Rhacochilus vacca (Pile perch)

<u>Materials</u>: 1 parasphenoid, 1 premaxilla, 1 lower pharyngeal.

<u>Total</u>: 3 specimens.

<u>Remarks</u>: The bones of the pile perch's jaw are diagnostic. Although more common today in Tacoma Narrows, there are a few reports of pile perch in Quartermaster Harbor (Miller and Borton 1980).

Order Pleuronectiformes (Left- and right-eyed flounders)

<u>Materials</u>: 4 articulars, 1 basihyal, 3 basypterygia, 3 ceratohyals, 1 cleithra, 3 dentaries, 1 epibranchial, 3 epihyal, 1 frontal, 3 hyomandibulae, 3 upper hypohyals, 1 lower hypohyal, 2 maxillae, 1 metapterygoid, 1

pharyngobranchial, 2 posttemporals, 1 premaxilla, 1 preopercle, 2 pterotics, 4 quadrates, 1 retroarticular, 1 scapula, 1 supracleithrum, 2 urohyals, 4 vomers, 3 1st vertebrae, 41 abdominal vertebrae, 96 caudal vertebrae, 1 hypural, 13 indeterminate vertebrae, 1 unidentifiable fragment.

<u>Total</u>: 205 specimens.

<u>Remarks</u>: There are two species of left-eyed flounder (family Bothidae) and 13 species of right-eyed flounder (family Pleuronectidae) found in Puget Sound today, and all have been reported either in Quartermaster Harbor or in the vicinity of Vashon Island. Specific flatfish taxa are diverse and difficult to determine from the fish bone in the Burton Acres Shell Midden assemblage. General rules about size (Pleuronectidae tend to be larger than Bothidae) do not apply here because of the range of size classes within this sample. Therefore most specimens that could be identified confidently to order Pleuronectiformes are left at that taxonomic level unless they are mouthparts or other well-preserved cranial elements. Pleuronectiformes are bottom-dwelling fish and have a long history of being used by humans. In this assemblage, pleuronectiformes are the most abundant type of fish after herring and salmon.

Family Pleuronectidae (Right-eyed flounders)

<u>Materials</u>: 2 ceratohyals.

<u>Total</u>: 2 specimens.

<u>Remarks</u>: These two cranial elements are identified to Pleuronectidae because they are definitely not from either Bothid species that inhabit Puget Sound. However, they cannot be firmly identified to any particular Pleuronectidae species.

cf. *Lepidopsetta bilineata* (Rock sole)

<u>Materials</u>: 1 frontal, 1 premaxilla.

<u>Total</u>: 2 specimens.

<u>Remarks</u>: Identification of these two elements as rock sole is based on exact matches with comparative material, but because the comparative collection is incomplete, this is a tentative identification. Puget Sound is a well-known spawning ground of the rock sole; this species has been reported within Quartermaster Harbor (Hart 1973; Miller and Borton 1980).

cf. *Microstomus pacificus* (Dover sole)

<u>Materials</u>: 1 premaxilla.

Table 10.2 Relative Abundance (NISP) of Fish Remains by Excavation Unit and Taxon

Unit	Layer	Volume Sampled (liters)	Dogfish C	C/V	Skate C	C/V	Ratfish C	C/V	Herring C	C/V	Salmon C	C/V	Toadfish C	C/V	Codfish C	C/V	Rockfish C	C/V	Sculpin C	C/V	Surfperch C	C/V	Flatfish C	C/V	Total C	C/V
22/58	2A‡	-	-	-	-	-	-	-	-	-	-	-	-	-	-	-	-	-	-	-	-	-	-	-	-	-
	2B‡	-	-	-	-	-	-	-	-	-	-	-	-	-	-	-	-	-	-	-	-	-	-	-	-	-
	2C‡	16	-	-	-	-	-	-	2	0.1	1	0.1	-	-	-	-	-	-	-	-	16	1.0	-	-	3	0.2
	2D	16	-	-	-	-	-	-	42	2.6	3	0.2	-	-	-	-	-	-	-	-	-	-	1	0.1	62	3.9
	2E	14	1	0.1	5	0.4	3	0.2	223	15.9	-	-	-	-	-	-	3	0.2	-	-	3	0.2	44	3.1	282	20.1
	2F	6	2	0.3	-	-	1	0.2	139	23.2	3	0.5	-	-	-	-	-	-	5	0.8	1	0.2	8	1.3	159	26.5
	2G	8	-	-	2	0.3	-	-	11	1.4	-	-	-	-	-	-	-	-	-	-	2	0.3	14	1.8	29	3.6
	3A	8	-	-	-	-	-	-	10	1.3	2	0.3	-	-	-	-	1	0.1	-	-	4	0.5	4	0.5	21	2.5
26/57	2A‡	26	-	-	-	-	-	-	1	0.1	-	-	-	-	-	-	-	-	-	-	-	-	-	-	1	0.1
	2B	10	-	-	-	-	-	-	2	0.2	-	-	-	-	-	-	-	-	-	-	-	-	-	-	2	0.2
	3A	4	-	-	-	-	-	-	1	0.3	-	-	-	-	-	-	-	-	-	-	-	-	1	0.3	2	0.5
28/58	2A‡	34	-	-	-	-	-	-	71	2.1	10	0.3	-	-	-	-	-	-	-	-	-	-	2	0.1	83	2.4
	2B‡	74	2	0.1	-	-	2	0.1	769	10.4	245	3.3	1	0.1	2	0.1	-	-	4	0.1	1	0.1	2	0.1	1028	13.9
	2C‡	32	2	0.1	1	0.1	1	0.1	149	4.7	34	1.1	-	-	-	-	-	-	-	-	1	0.1	11	0.3	199	6.2
	3A	16	-	-	-	-	1	0.1	22	1.4	2	0.1	-	-	-	-	-	-	-	-	-	-	6	0.4	31	1.9
29/58	2A‡	40	1	0.1	-	-	-	-	160	4.0	6	0.2	3	0.1	1	0.1	1	0.1	-	-	-	-	1	0.1	168	4.2
	2B‡	22	-	-	-	-	-	-	176	8.0	16	0.7	-	-	3	0.1	5	0.1	-	-	-	-	4	0.2	201	9.1
	2C‡	44	1	0.1	-	-	-	-	50	1.1	76	1.7	-	-	-	-	8	0.3	2	0.1	4	0.1	27	0.6	168	3.8
	2D	32	2	0.1	8	0.3	3	0.1	136	4.3	46	1.4	-	-	-	-	4	0.1	4	0.1	9	0.3	11	0.3	227	7.1
	2E	36	3	0.1	9	0.3	4	0.1	984	27.3	94	2.6	2	0.1	1	0.1	4	0.1	24	0.7	28	0.8	36	1.0	1189	33.0
	2F	24	1	0.1	5	0.2	2	0.1	1332	55.5	30	1.3	-	-	1	0.1	6	0.3	20	0.8	23	1.0	46	1.9	1466	61.1
Total		462	15	0.1	30	0.1	17	0.1	4280	9.3	568	1.2	6	0.1	8	0.1	28	0.1	59	0.1	92	0.2	218	0.5	5321	11.5

‡ = layers thought to date to post-contact with Euro-Americans; all other layers date to pre-contact (see Table 5.1). C = count. C/V = count/volume.

Total: 1 specimen.

Remarks: This identification is tentative because of the incomplete comparative collection used. The Dover sole is considered one of the most hardy flatfishes, living off the bottom of waters after an extended pelagic (living free from the bottom) life as a juvenile (Hart 1973). Dover sole have been reported in Quartermaster Harbor and surrounding Vashon Island (Miller and Borton 1980).

cf. *Parophrys vetulus* (English sole)

Materials: 1 premaxilla, 1 quadrate, 1 vomer.

Total: 3 specimens.

Remarks: Identification of this taxon is based on exact matches, but like the other species-level identifications of flounder taxa it must remain tentative because of limited comparative material. English sole congregate seasonally in shallow water in the spring shortly after spawning, and move into deeper waters by winter (Hart 1973). Like other species of flatfish, English sole is common in the Puget Sound and has been reported in Quartermaster Harbor (Miller and Borton 1980).

cf. *Platichthys stellatus* (Starry flounder)

Materials: 1 premaxilla, 2 interhaemal spines.

Total: 3 specimens.

Remarks: Starry flounder are mainly a shallow water flatfish and are unique among Pleuronectidae in that they may be either left-eyed or right-eyed. They also have an uncommon tolerance for freshwater (for flatfish), and they have been found in larger rivers emptying into the Pacific Ocean (Hart 1973). Starry flounder are a common fish both in the Puget Sound area and around Vashon Island specifically (Miller and Borton 1980).

RESULTS

Almost 9000 fish remains representing 21 taxa are identified from the Burton Acres Shell Midden, 5321 (about 60%) of which are identified to at least order-level. Clupeidae (herring) dominates the assemblage as a whole, representing approximately 80% of the identified specimens. Salmonid remains constitute approximately 10% of the fish fauna, Pleuronectiformes (flatfish) provide about 4% of the remains, and the remaining 18 taxa make up the balance. About 3500 fish remains cannot be identified to any particular

taxon. The abundances of fish bone divided into different taxa are given in Table 10.2. This table lists the four excavation units divided into their respective stratigraphic layers. The volume of excavated sediment in liters from each layer that was sampled for the analysis of fish bone is also listed. For each taxa, two numbers are given. The first is a count of all specimens belonging to that taxa in a particular unit and layer. The second is the count divided by the number of liters sampled in the layer in which it was found. The count divided by volume of the sample standardizes the counts, and gives a measure of density that can be compared horizontally and vertically across the site.

Although most fish taxa are seen in small densities throughout all units and layers, with the exception of Unit 26/57, there are some noticeable patterns in the distribution of some taxa. Corrected for volume, Table 10.2 shows that the densest deposits of fish bones in Unit 22/58 are in the middle layers, 2E and 2F. Herring are the only substantial taxa in this unit, although dogfish, skate, ratfish, salmon, rockfish, sculpin, surfperch, and flatfish are present in small quantities. There are very few fish bones in Unit 26/57, only four herring and one flatfish specimen. The densest deposits of fish bones in Unit 28/58 are in the middle layers, 2B and 2C. Herring and, to a much lesser extent, salmon are the dominant taxa here. Small numbers of dogfish, skate, ratfish, toadfish, cod, sculpin, surfperch, and flatfish are present as well. Unit 29/58 has the densest deposits of fish bone in its bottom layers, 2E and 2F. Like the other units, this one is dominated by herring and salmon, but contains numerous remains of flatfish, surfperch, and sculpin. Dogfish, skate, ratfish, toadfish, cod, and rockfish are also present.

DISCUSSION

Taphonomic Summary

The fish bones at 45KI437 are well preserved, which is common in shell middens because of the alkalinity of the surrounding sediment. Most soils in the Northwest are acidic rather than alkaline, due to the decomposition of organic matter and release of organic acids. The calcium carbonate in the shells of the midden turns the sediment alkaline and lessens the effects of these acids. The result of this alkalinity at the Burton Acres Shell Midden is a fish bone assemblage that is well preserved. Very delicate herring cranial fragments

Table 10.3 Relative Abundance (NISP) of Burned Fish Remains by Excavation Unit and Taxon

Unit	Layer	Volume Sampled (liters)	Dogfish C	Dogfish C/V	Ratfish C	Ratfish C/V	Herring C	Herring C/V	Salmon C	Salmon C/V	Toadfish C	Toadfish C/V	Codfish C	Codfish C/V	Rockfish C	Rockfish C/V	Surfperch C	Surfperch C/V	Flatfish C	Flatfish C/V	Unid.Taxa C	Unid.Taxa C/V	Total C	Total C/V
22/58	2A‡	0	-	-	-	-	-	-	-	-	-	-	-	-	-	-	-	-	-	-	-	-	-	-
	2B‡	0	-	-	-	-	-	-	-	-	-	-	-	-	-	-	-	-	-	-	-	-	-	-
	2C‡	16	-	-	-	-	-	-	1	0.1	-	-	-	-	-	-	-	-	-	-	-	-	1	0.1
	2D	16	-	-	-	-	-	-	1	0.1	-	-	-	-	-	-	-	-	-	-	14	0.9	15	0.9
	2E	14	-	-	-	-	-	-	-	-	-	-	-	-	-	-	-	-	-	-	-	-	-	-
	2F	6	-	-	-	-	-	-	-	-	-	-	-	-	-	-	-	-	-	-	-	-	-	-
	2G	8	-	-	-	-	-	-	-	-	-	-	-	-	-	-	-	-	-	-	-	-	-	-
	3A	8	-	-	-	-	-	-	-	-	-	-	-	-	-	-	-	-	-	-	-	-	-	-
26/57	2A‡	26	-	-	-	-	-	-	-	-	-	-	-	-	-	-	-	-	-	-	-	-	-	-
	2B	10	-	-	-	-	-	-	-	-	-	-	-	-	-	-	-	-	-	-	-	-	-	-
	3A	4	-	-	-	-	-	-	-	-	-	-	-	-	-	-	-	-	-	-	-	-	-	-
28/58	2A‡	34	-	-	-	-	-	-	2	0.1	-	-	-	-	-	-	-	-	-	-	-	-	2	0.1
	2B‡	74	-	-	1	0.1	17	0.2	219	3.0	-	-	-	-	-	-	-	-	-	-	120	1.6	357	4.8
	2C‡	32	-	-	1	0.1	79	2.5	25	0.8	-	-	-	-	-	-	1	0.1	-	-	56	1.8	162	5.1
	3A	16	-	-	1	0.1	-	-	-	-	-	-	-	-	-	-	-	-	-	-	-	-	1	0.1
29/58	2A‡	40	-	-	-	-	-	-	-	-	-	-	-	-	-	-	-	-	-	-	-	-	-	-
	2B‡	22	-	-	-	-	21	1.0	11	0.5	1	0.1	-	-	-	-	-	-	2	0.1	3	0.1	38	1.7
	2C‡	44	-	-	-	-	25	0.6	58	1.3	-	-	1	0.1	1	0.1	-	-	7	0.2	54	1.2	146	3.3
	2D	32	-	-	-	-	-	-	-	-	-	-	-	-	2	0.1	-	-	1	0.1	-	-	3	0.1
	2E	36	1	0.1	1	0.1	15	0.4	5	0.1	-	-	-	-	-	-	1	0.1	-	-	-	-	23	0.6
	2F	24	1	0.1	-	-	78	3.3	7	0.3	-	-	-	-	-	-	8	0.3	5	0.2	5	0.2	104	4.3
Total		462	2	0.1	4	0.1	235	0.5	329	0.7	1	0.1	1	0.1	3	0.1	10	0.1	15	0.1	252	0.5	852	1.8

‡ = layers thought to date to post-contact with Euro-Americans; all other layers date to pre-contact (see Table 5.1). C = count. C/V = count/volume.

are preserved to the point of being identified to particular element. Not all shell middens in the Northwest have such excellent preservation.

Burned fish bones are present in small quantities in certain parts of the site (see Table 10.3). Lyman (1994) notes that color is the most frequently used criterion for determining whether or not bone has been burned. Shipman et al. (1984) performed numerous experimental studies on this subject and caution against the use of color to judge the nature of burning on bone, as numerous chemical processes during bone diagenesis can change the color of bone in a similar manner. Instead, proper analysis of bone for evidence of burning, and the temperature range to which bone has been exposed, can be done effectively by examining the specimen's crystalline structure using scanning electron microscopy (SEM). The difference between carbonized bone, which has been blackened, and calcined bone, which has turned white and lost all organic material in its chemical composition, is the temperature to which it has been exposed. McCutcheon (1992) performed experimental studies as well on the chemical and microscopic changes that occur to bones at high temperatures.

Because SEM and other techniques are quite expensive, and the small number of elements in this assemblage whose colors are characteristic of burning, this analysis used color as sufficient for recording whether or not a fish bone has been "burned". The summary of burned bone is given in Table 10.3. Within each column, the first number is the count of burned bone for each taxa in each unit and layer. The second number is the standardized count per volume. The last column gives the total abundance of burned fish bone in each unit and layer.

Of all the fish bones analyzed, 852 were identified as burned. Most burned fish bone is found in Units 28/58 and 29/58. In Unit 22/58, 16 burned specimens were found, and no burned specimens were found in Unit 26/57. The most notable concentration of burned fish bone is in Unit 28/58, Layer 2B. A dense deposit of salmon bones and other unidentifiable fish bones were found here, burned to the point of grayish-white discoloration, possibly indicating high temperatures. A deposit of burned herring bones lies below this in Layer 2C. Unit 29/58 had fewer burnt bones than Unit 28/58, and these tend to occur in lower layers than the burned bones in Unit 28/58. A small concentration of burned

salmon, herring, and unidentifiable fish bones occurs in Layer 2C, and a small concentration of burned herring bones occurs in Layer 2F of Unit 29/58. Other taxa display evidence of burning, including dogfish, ratfish, flatfish, codfish, surfperch, and rockfish. They occur in small numbers throughout Units 28/58 and 29/58.

The burned fish bones at 45KI437 are evidence of fire use at Burton Acres Shell Midden. Possible fires associated with a shell midden such as Burton Acres Shell Midden are cooking hearths and food drying operations, as well as fires of natural origin. Bones could have been burned as they fell off a rack and into a fire used to smoke and dry fish.

Markings are more difficult to interpret on fish bones than coloration indicating burning. These marks, such as butchering and cut marks, are usually rare on fish remains and are not present on any of the remains analyzed from Burton Acres Shell Midden.

Differential survivorship of bone fragments due to bone density has been documented as an important factor in fish bone assemblage formation (Butler 1987, 1993). There are few salmon cranial elements as opposed to vertebrae, which in many contexts can be accounted for by bone density. However, there is an abundance of herring cranial fragments at 45KI437, which are even less dense. Therefore, bone density probably played a minor role in the composition of this assemblage.

Effects of Screen Size

Fish bones, like the fish they come from, vary greatly in size. The Pacific halibut (*Hippoglossus stenolepis*), is one of the largest bony fishes in the north Pacific, sometimes reaching over 2 1/2 meters in length, while the smallest species of fish are only a few centimeters long. The size of the fish and their fragmentation after death affect the screen size distribution of the assemblage. Although large fish such as Pacific halibut have been found at the mouth of Quartermaster Harbor (Miller and Borton 1980), the largest species found in the fish bone assemblage is the cabezon (*Scorpaenichthys marmoratus*). Most of the specimens found were Pacific herring (*Clupea harengus pallasi*), or fragmented salmon (*Oncorhyncus sp.*).

The screen size distribution reflects the composition of taxa: almost all specimens are from either 1/4 inch or 1/8 inch screens (Figure 10.1). For example, in Units 28/58 and 29/58, 5.4% of the specimens are from the 1/4

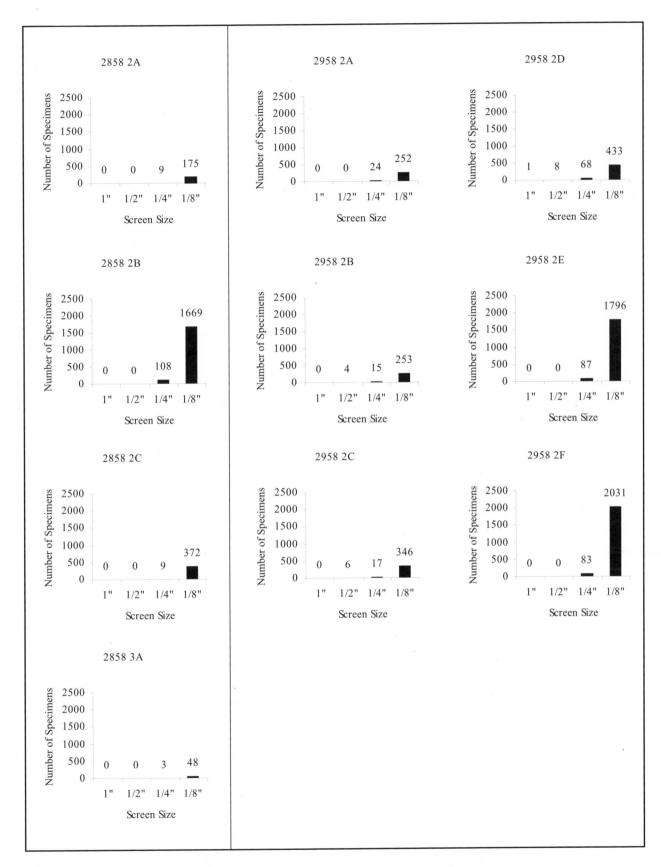

Figure 10.1 Number of fish remains found in each screen size displayed by layer and excavation unit. Numbers above bars represent counts.

Figure 10.2 Number of herring, salmon, and other kinds of fish in each layer of Units 28/58 and 29/58 (combined). Numbers in parentheses represent NISP of herring/salmon/other.

inch screen, while 94.3% of the specimens are from the 1/8 inch screen. There was only one specimen in the 1 inch fraction: a cabezon (*Scorpaenichthys marmoratus*) fragment in Unit 29/58, Layer 2D. There were 20 specimens in the 1/2 inch fraction, 18 of which were in Layers 2B, 2C, and 2D of Unit 29/58. They were large herring cranial elements, salmon vertebrae, and six specimens that may be part of the cabezon, as they were found in the same bucket, but cannot be identified positively to that taxon. The other two specimens found in the 1/2 inch mesh were a flatfish (*Pleuronectiformes*) posttemporal and a rockfish (*Scorpaenidae*) abdominal vertebra found in Unit 22/58 Layer 2E (not shown in Figure 10.1). The majority of 1/8 inch specimens is not surprising given the abundance of taxa that are of small sizes (e.g. herring, surfperch).

Variability Over Time

Units 28/58 and 29/58 at the Burton Acres Shell Midden contain material spanning perhaps the last thousand years (see Chapter 5). Although the two other excavation units contain fish bone, their sample sizes are much smaller than these two units. Also, Units 28/58 and 29/58 are adjacent and have comparable stratigraphy, therefore the fish bone composition in these units will be analyzed for change over time.

The change in fish bone taxonomic composition can be seen stratigraphically in these two units (Figure 10.2) by examining the changing proportion of herring, salmon, and other fish taxa. These three categories are represented proportionally in each bar. The bars represent the fish bone assemblage sampled in each layer of Units 28/58 and 29/58 combined. Layer 2A is stratigraphically the younger layer.

The graph shows that herring is consistently the most abundant taxa in each layer, representing approximately 50% of the assemblage in Layer 2C to over 90% in Layer 2A. Salmon usually comprises the second most abundant category. It is proportionally the most abundant in Layer 2C, when herring is at its minimum. The other fish taxa found at the site, combined into one analytic category, comprise over 20% of the assemblage in Layer 2D and almost 20% in Layer 2C, when salmon is relatively abundant and herring is less abundant. In the densest layers, 2E and 2F, herring is maximally abundant, and salmon is less abundant than the other taxa found at the site. Although these other taxa were not differentiated in the graph, there is a marked increase in flatfish and surfperch in these layers (see Table 10.2). The layers towards the bottom are dominated by herring, despite an increase in other taxa. This increase in the number of other taxa found is most likely the result of the larger size of the assemblages towards the bottom. Grayson (1984) notes that greater taxa richness can be expected with larger sample sizes.

Based on the fish taxa composition shown in Figure 10.2, there is a change stratigraphically in the fish being deposited at the Burton Acres Shell Midden. Herring were being deposited in much greater quantities in the bottom of these units, in layers associated with denser shells and shell fragments (see Chapter 11). The upper layers of Units 28/58 and 29/58, containing fewer fish bones, still contain proportionally more herring specimens than other taxa, although salmon is present in higher numbers in these upper layers relative to the lower layers.

CONCLUSIONS

Fish bones are the second most abundant type of faunal remains, next to shell, at Burton Acres Shell Midden. A sample of slightly over 25% of the fish bone at the site is analyzed. Almost 9000 individual specimens are examined. Of these, 5326 (approximately 60%) are identified taxonomically to at least family level (or order, in the case of some flatfish). They represent 20 fish taxa, all of which are common in the Quartermaster Harbor and Vashon Island area (Miller and Borton 1980).

Most of the fish bone came from the 1/8 inch screen size fraction, which is not surprising due to the abundance of herring in the assemblage. One element was found in the 1 inch fraction, belonging to cabezon, the largest taxa found at 45KI437. Based on the presence of herring cranial elements, the bones seem to be preserved quite well, a common occurrence at shell middens. Slightly less than 10% of the fish bone was burned, mostly salmon remains in Unit 28/58.

The fish bone assemblage at 45KI437 provides a record of changing patterns in fish use on Vashon Island. Fish utilization at Burton Acres Shell Midden included a variety of fish commonly available in the local area and seasonally restricted resources such as herring and salmon. Herring was the most abundant fish taxa, especially in the lower layers of the site. Although herring continued to be the most abundant taxa over time, the upper layers, most notably 2D, 2C, and 2B associated with historic artifacts exhibit a relatively greater proportion of salmon and other taxa.

REFERENCES

Butler, V.L.

1987 Distinguishing Natural from Cultural Salmonid Deposits in the Pacific Northwest of North America. In *Natural Formation Processes and the Archaeological Record*, edited by D.T. Nash and M.D. Petraglia, pp. 131-149. British Archaeological Reports International Series No. 352.

1993 Natural Versus Cultural Salmonid Remains: Origin of the Dalles Roadcut Bones, Columbia River, Oregon, U.S.A. *Journal of Archaeological Science* 20:1-24.

Cannon, D.Y.

1987 *Marine Fish Osteology, A Manual for Archaeologists*. Publication No. 18. Department of Archaeology, Simon Fraser University, Burnaby, British Columbia.

Grayson, D.K.

1984 *Quantitative Zooarchaeology*. Academic Press, New York.

Hart, J.L.

1973 *Pacific Fishes of Canada*. Bulletin No. 180. Fisheries Research Board of Canada.

Lyman, R.L.

1994 *Vertebrate Taphonomy*. Cambridge University Press, Cambridge.

McCutcheon, P.T.

1992 Burned Archaeological Bone. In *Deciphering a Shell Midden*, edited by J.K. Stein, pp. 347-370. Academic Press, New York.

Miller, B.S., and S.F. Borton

1980 *Geographical Distribution of Puget Sound Fishes, Maps and Data Source Sheets*. Fisheries Research Institute, University of Washington, Seattle.

Quinn, T.P., B.S. Miller and R.C. Wingert

1980 Depth Distribution and Seasonal and Diet Movements of Ratfish, *Hydrolagus collei*, in Puget Sound, Washington. *Fisheries Bulletin* 78:816-821.

Shipman, P., G. Foster and M. Schoeninger

1984 Burnt Bones and Teeth: An Experimental Study of Color, Morphology, Crystal Structure and Shrinkage. *Journal of Archaeological Science* 11:307-325.

Wydoski, R.S., and R.R. Whitney

1979 *Inland Fishes of Washington*. University of Washington Press, Seattle.

11

Faunal Analysis: Shellfish Remains

Laura S. Phillips

Shell middens are, of course, characterized by an abundance of shell. Burton Acres Shell Midden is no exception. The intertidal zone at Burton Acres provided a wonderful abundance of shellfish, much of which could be eaten freshly-steamed. Ethnographically, cockles and Horse clams were harvested and dried by the S'Homamish in the spring or summer to preserve them for winter food. The windy point at Burton Acres Shell Midden would have provided an excellent drying spot for just such a job. Littleneck clams and Butter clams were by far the most abundant shellfish in this archaeological site, although a wide variety of other shellfish species appear in smaller quantities.

As in most shell middens in Puget Sound, the abundant fragments of shellfish remains at the Burton Acres Shell Midden reflect the importance of shellfish in subsistence. Analysis of the shellfish recovered from the site indicates that people were procuring and utilizing shellfish from the intertidal zone of Quartermaster Harbor and a few other areas. Shells are ubiquitous in shell middens, but are typically fragmented and can be difficult to identify to species, genus, or even family level. Fortunately, the preservation of shell and other fauna is excellent at this site, and a wide variety of shellfish types could be identified to genus or species level.

METHODOLOGY

Sampling

To gather data about the shellfish in the midden, a sampling strategy was developed to collect an adequate and spatially-even sample in the field. All of the shells recovered from the 1 inch and 1/2 inch size screens were collected, as were more than 25% of the 1/4 inch and 1/8 inch shell (the first bucket in each layer, and every bucket divisible by four). All of these shells were cataloged and curated.

One hundred percent of the samples saved were selected for analysis from the 1 inch, 1/2 inch, and 1/4 inch size screens for 15 of the 21 excavated layers. The remaining six layers were more voluminous, and large samples of shell had been collected. For these six layers fewer samples had to be analyzed to achieve redundancy (the point at which, statistically, the number of new species identified does not increase appreciably). In all, over 23% of the total shell (by volume) was analyzed (see Appendix I).

The shell from the 1/8 inch mesh-size screen was weighed, but not taxonomically differentiated for analysis, based on an initial analysis from Unit 22/58. In order to determine whether or not more data could be

---

Chapter opening photo: Archaeologist Bob Kopperl screens a bag from excavation. Almost every bucket removed from the site had shellfish remains in it, and screening the bucket contents was the only way to see that shell. Every fourth bucket was sorted entirely, taking as long as five hours to separate the shell from the other constituents.

gained by identifying the 1/8 inch shell fragments, a portion of the shell from the 1/8 inch size screens from Unit 22/58 was analyzed taxonomically. In Table 11.1, the abundance of shellfish taxa identified in this 1/8 inch sample is compared to the abundance of shellfish taxa from Unit 22/58 in the other mesh screens.

Table 11.1 shows that no new taxa were identified in the 1/8 inch sample. And with the exception of a few hinges of *Protothaca staminea* (Pacific Littleneck Clam), the 1/8 inch fragments of shell could not be identified more specifically than to genus. Finer taxa in the 1/8 inch sample could not be identified because the fragments lacked distinguishing features such as whole hinges or columella. As such, the volume of shell fragments identified to the grossest taxon, unidentified Mollusca, increases in abundance, while more specific identification of taxa such as Veneridae and Bivalvia decrease in abundance. The abundance of barnacle, mussel, and cockle is likely artificially inflated in comparison to the other taxa because these invertebrates have distinctive characteristics that are easily identifiable even in the 1/8 inch size-fraction.

The absence of more specific taxa and the decrease in the abundance of the taxa identified in this 1/8 inch sample suggests that the 1/8 inch shellfish fragments are merely smaller fragments of those shellfish sampled in the larger mesh screens. Based on these data, coupled with time constraints, no other shellfish from the 1/8 inch mesh-size screen was taxonomically differentiated.

Sorting

Shellfish remains with the same provenience (unit, layer, bucket, screen-size) were sorted by taxonomic class. Each class was put in a separate bag containing an identification label. A comparative reference collection comprised of contemporary shells collected from the site was used for the analysis, as well as the archaeological comparative collection collected from the British Camp site (45SJ24, Op. A) on San Juan Island and analyzed by Pamela J. Ford (1992). Additionally, the malacology collection housed at the Burke Museum served as a reference for the freshwater species. A portion of the contemporary reference collection was donated by King County biologists who were studying shellfish populations in Quartermaster Harbor as part of the 1996 Beach Steward Program.

Analysis Protocol

In addition to taxonomic identification, each specimen was described and inspected for burning or other modification. Identifications and modifications were recorded and entered into a database (see Appendix H).

DESCRIPTIVE SUMMARY

This section provides a descriptive summary of the identified invertebrate faunal specimens by taxon. Taxonomic range and habitat are noted as well as the total weight and Minimum Number of Individual specimens (MNI) for each taxonomic category. Classifications follow Abbott (1974) and Kozloff (1996).

Phylum Mollusca (Molluscs)
<u>Material</u>: indistinct valve fragments
<u>Total</u>: 2915g
<u>Remarks</u>: These specimens could only be identified as the hard calcium carbonate valves of soft-bodied invertebrates. No distinguishing features could be identified. Approximately 23% of the total shell was identified to this level.

Class Bivalvia (Bivalves)
<u>Material</u>: indistinct valve and hinge fragments
<u>Total</u>: 2354.7g (MNI - 40)
<u>Remarks</u>: Bivalves are molluscs with pairs of hinged valves. The specimens in this category include all unidentified bivalves. The majority of these specimens were probably Venus clams and cockles from which distinguishing features were absent or had eroded.

Order Mytiloida
Family Mytilidae
Mytilus spp. (Blue Mussels)
<u>Material</u>: valve and hinge fragments
<u>Total</u>: 113.1g (MNI - 59)
<u>Remarks</u>: In Family Mytilidae, there are four species common to the Northwest: *Mytilus edulis*, *Mytilus californianus*, *Modiolus rectus*, and *Modiolus flabellatus* (considered a subspecies of *Modiolus rectus* by some). The *Mytilus* species exploit a rocky substrate near shore, and the *Modiolus* species tend towards a moderately deepwater habitat. In this site, only hinges of *Mytilus* spp. were identified. Due to specimen size of the valves (*M. edulis* tend to be smaller than *M. californianus*), the habitat found in the direct vicinity of Burton Acres Shell Midden, and the lack of any *Modio-*

lus spp. hinges or valve fragments identified, all mussel fragments were identified as *Mytilus* spp.

Order Pterioida
Family Ostreidae (Oysters)
Material: valve and hinge fragments
Total: 1.5g
Remarks: *Ostrea lurida*, which prefers intertidal zones with sand or gravel, is the only oyster native to the Northwest Coast. The two other possible species *Crassostrea gigas* and *Crassostrea virginica*, however, were introduced to the Northwest Coast in the early part of this century (Carlton in Ford 1995). Oyster shells are difficult to identify to species because of the layered structure of the shell, which fragments easily.

Family Pectinidae
Patinopecten caurinus (Giant Pacific Scallop)
Material: valve fragments
Total: 5g
Remarks: These edible scallops inhabit sandy or muddy bottoms, in depths from 10 to 200 meters. Although the site lies in the shallow waters of inner Quartermaster Harbor, the waters are deep near the entrance to Quartermaster Harbor. These specimens were most likely procured from the beach where the shell would have washed ashore after storms (Wessen 1994b:337). Few of these specimens have been recovered from sites in southern Puget Sound. However, Harlan Smith (1907:395) mentions their presence in the shell middens of Quartermaster Harbor. In addition, Alan Bryan (1963:Appendix B) notes that two specimens were found at each Dugualla Bay (45IS35) and Penn Cove Park (45IS50) in Island County. The few fragments found in the Burton Acres Shell Midden assemblage appear to be ground. Ethnographically, these scallop shell were used as decorative objects and rattles (Stewart 1973:162,165).

Chlamys rubida (Hind's or Pink Scallop)
Material: whole valve
Total: 0.8g (MNI - 1)
Remarks: This scallop is common from depths of five meters in subtidal zones with strong currents, but can be found in waters as deep as 200 meters. It is rarely found in archaeological sites throughout Puget Sound. This single specimen has not been culturally modified.

Order Veneroida
Family Cardiidae
Clinocardium spp. (Cockles)
Material: valve and hinge fragments
Total: 351.2g (MNI - 23)
Remarks: Cockles are common to Puget Sound, and are most often found in soft, sand, gravel or mud substrates in high intertidal zones or shallow subtidal zones. They are common in archaeological sites throughout Puget Sound. The only species of cockle identified at the Burton Acres Shell Midden is *Clinocardium nuttalli*. Other cockle species could be present, including *Clinocardium ciliatum* and *Clinocardium fucanum*, but no hinges of this species were identified, so all cockles valves were attributed to *C. nuttalli*.

Clinocardium nuttalli (Basket Cockle)
Material: valve and hinge fragments
Total: 190.3g (MNI - 1)
Remarks: Basket cockles have valves with distinctive coarse radial ribs often with moon-shaped riblets (Abbott 1974:487). They are reported in nearly every archaeological shell midden site in Puget Sound.

Family Veneridae (Venus Clams)
Material: valve and hinge fragments
Total: 4554.3g (MNI - 54)
Remarks: This family includes species that are prolific to most shell middens in Puget Sound, *Protothaca staminea* (Native Littleneck clam) and *Saxidomus giganteus* (Butter clam). These species typically exploit the intertidal zone. The introduced species of Littleneck, the Japanese Littleneck (*Tapes philippinarum*) is similar in appearance to the native species, but exploits the higher intertidal zones, and does not appear to compete with the Native Littleneck for habitat. Since the radial and transverse markings on Littleneck clams of several genera are distinctive, most of these fragments could be identified to family level. Only two species were identified in the site from this family: *P. staminea* and *S. giganteus*. Since valve fragments without hinges of *S. giganteus* can only be identified to class, a majority of the fragments identified to the Veneridae family are most likely *P. staminea*.

Protothaca staminea (Native Littleneck Clam)
Material: whole valves, valve fragments with hinges

Total: 847.7g (MNI - 338)

Remarks: Due to similarities between Venus clams, only hinges and valves with nearly complete hinges were identified to the species, *Protothaca staminea*. This species is common to many shell middens in Puget Sound.

Saxidomus spp.

Material: valve fragments with hinges

Total: 262.7g (MNI - 45)

Remarks: Two species of *Saxidomus* are found in the lower portions of intertidal waters of the Northwest Coast, *Saxidomus nuttalli* and S*axidomus giganteus*. These species are quite similar, but prominent concentric ridges on the valve distinguish *S. nuttalli*.

Saxidomus giganteus (Butter or Washington Clam)

Material: valve fragments with hinges

Total: 19.9g (MNI - 7)

Remarks: Butter clams typically inhabit sand, shell, mud, or gravel bottoms in the lower third of the intertidal zone to 15 meters in the subtidal. They are the largest of the Venus clams in this site. Unlike the Littleneck clams, Butter clams have no radial markings. As such, without a hinge or portion of a hinge, these specimens can only be identified as Bivalvia.

Family Mactridae
Tresus spp. (Horse and Gaper Clams)

Material: valve fragments with hinges

Total: 357g (MNI - 49)

Remarks: *Tresus capax* and *Tresus gapperi* inhabit sheltered sand or mud substrates in the lower third of the intertidal zone to as deep as 20 meters in the subtidal zone. The two species can be distinguished by the outline shape of the valve, and require whole or nearly whole specimens for adequate identification. No whole valves were recovered from the site.

Family Tellinidae
Macoma nasuta (Bent-nosed Clam)

Material: valve fragments with hinges

Total: 2.1g (MNI - 5)

Remarks: This species of clam is fragile and the distinguishing feature of the shell erodes quickly. Few specimens were identified in this site, but may be underrepresented due to their tendency to fragment and erode (Ford 1995).

Order Myoida
Family Myidae
Mya arenaria (Softshell Clam)

Material: valve fragment with hinge

Total: 3.9g (MNI - 1)

Remarks: This species is not native to the Northwest, and was introduced in the 1880s from the Atlantic Coast of the U.S. (Cohen and Carlton 1995). It is common in sandy mud estuaries. A single hinge was found in the uppermost layer of Unit 29/58. The historic date associated with the uppermost Layer, 2A, is consistent with the introduction of this species.

Order Arcoida
Family Glycymerididae
Glycymeris subobsoleta (Pacific Coast Glycymeris)

Material: valve with hinge

Total: 4.8g (MNI - 1)

Remarks: The presence of taxodont dentition (numerous teeth on the hinge) along with raised ribs and interior crenulations on the ventral border, distinguish this species. A single specimen was found in an unsampled bucket (Unit 29/58, Layer 2F, Bucket 5) and, therefore, is not part of this analysis. However, this specimen is worthy of note because of its relative absence at other sites in Puget Sound. Specimens have been reported on the outer coast of Washington at the Ozette site (45CA24) (Wessen 1994b:334). Wessen suggests that the shell was collected incidentally during the procurement of *Protothaca staminea* and *Saxidomus giganteus*.

Order Schizodonta
Family Unionidae
Unio margartifera (Freshwater Clam)

Material: valve with hinge

Total: 12.8g (MNI - 1)

Remarks: A nearly whole valve of this freshwater specimen was identified using the comparative malacology collections at the Burke Museum. The closest freshwater source to the site is Judd Creek, and at one time, could have been large enough to support a freshwater shellfish population. This single specimen could have been deposited in the midden in several ways. It could have been brought as a trade item, procured at a larger freshwater source on or off Vashon Island, specifically procured along the creek, or accidentally

Table 11.1 Abundance of Shellfish Taxa (grams/liter) in Unit 22/58: Comparison of 1 inch, 1/2 inch, and 1/4 inch Screen, with 1/8 inch Screen

Unit	Layer	Volume Sampled (liters)	Mollusca														Arthropoda		Total Shell Volume (g/l)	
			Bivalvia											Unidentified Mollusc		Balanus spp. (Barnacle)			
			Cardiidae Clinocardium spp. (Cockle)		Mytilidae Mytilus spp. (Mussel)		Veneridae (Venus clam)				Unidentified Bivalve								
							Prothaca staminea (Littleneck clam)		Unidentified Venus clam										
			≥1/4"	1/8"	≥1/4"	1/8"	≥1/4"	1/8"	≥1/4"	1/8"	≥1/4"	1/8"	≥1/4"	1/8"	≥1/4"	1/8"	≥1/4"	1/8"
2258	2A‡	10	-	-	-	-	-	-	0.1	-	0.2	-	0.1	0.2	-	-	0.4	0.3
	2B‡	36	-	-	-	-	-	-	0.2	-	0.1	-	-	-	-	-	0.4	-
	2C‡	16	-	-	-	-	-	-	0.6	-	0.3	0.1	0.3	0.5	-	0.1	1.2	0.6
	2D	12	0.4	0.4	0.1	0.2	0.4	0.1	4.0	0.2	2.2	0.4	3.3	5.8	0.5	0.5	12.0	7.6

‡ = layers thought to date to post-contact with Euro-Americans; all other layers date to pre-contact (see Table 5.1).
≥1/4" = shell found on 1 inch, 1/2 inch, and 1/4 inch screen (combined); 1/8" = shell found on 1/8 inch screen.

collected near the mouth of the creek during marine shellfish procurement activities. The interiors of these freshwater clams are distinctly pearled, and could have been used as decoration.

The single valve was recovered from the lowest layer of Unit 29/58. This layer was inundated by water during high tide, and, thus, has better preservation of organic materials than dry layers. A freshwater shellfish (a single *Margaritana margaritifera*) has only been recorded at one other archaeological site in Puget Sound, Penn Cove Park (45IS50) (Bryan 1963:Appendix B:9) in Island County. Like the Burton Acres Shell Midden, a portion of Penn Cove Park was also inundated at high tide, and, thus, could have created better preservation of such fragile shellfish. At the Ozette site, where there was also excellent preservation, a single specimen (*Margaritana margaritifera*) was reported (Wessen 1994a:129).

Class Gastropoda (Snails and Slugs)

Material: fragments, columellae

Total: 10.4g (MNI - 4)

Remarks: These univalve specimens could not be identified to finer taxonomic levels due to the lack of characteristic features on columellae and fragments. The most common gastropods found in archaeological sites in the Puget Sound region include limpets, periwinkles, whelks, and moonsnails.

Order Patellogastropoda
Family Lottidae (Limpets)

Material: fragments, fragments with apex

Total: 0.3g

Remarks: All limpets are included in this family except for *Acmaea mitra*, which belongs to Family Acmaeidae. Limpets exploit rocky substrates in the middle and upper intertidal zones. No limpets were identified to finer taxonomic level due to their fragmentary nature in the site. The few specimens recovered from this site were found in the lower layers of Units 28/58 and 29/58.

Order Mesogastropoda
Family Littorinidae (Periwinkles)

Material: body whorl or spine fragments, columellae

Total: 1.9g (MNI - 4)

Remarks: Periwinkles prefer rocky substrates in the intertidal zone, among barnacles and mussels. Common species found in archaeological sites in Puget Sound include *Littorina scutulata* and *Littorina sitkana*. The few specimens recovered from this site were found in the units closest to the water's edge, Units 28/58 and 29/58.

Family Naticidae
Polinices spp. (Moon Snails)

Material: body whorl or spine fragments, columellae

Total: 97.7g

Remarks: Moon snails have robust shells and can grow up to 14cm in height. During the summer months, they are typically found in sandflats in the intertidal zone, but exploit the subtidal as well. The most common member of this species found in Puget Sound archaeological sites is *Polinices lewisii*, Lewis' Moon Snail.

Order Neogastropoda
Family Nucellidae
Nucella spp. (Dogwinkles)

Material: body whorl or spine fragments, columellae

Total: 12.7g (MNI - 2)

Remarks: Dogwinkles inhabit the rocky substrates of the middle and upper intertidal zone in order to prey on barnacles. The two most common species recovered archaeologically from this genus include *Nucella emarginata* and *Nucella lamellosa*.

Family Nassariidae (Dog Whelks)

Material: body whorl or spine fragments, columellae

Total: 1.3g (MNI - 5)

Remarks: Members of this family inhabit mud substrates in the intertidal zone. The most common native species in Puget Sound include *Nassarius fossatus*, *Nassarius mendicus*, and *Nassarius perpinguis*.

Class Polyplacophora (Chitons)

Material: plate

Total: 0.1g

Remarks: Chiton shells are made up of a series of shelly plates. They cling to rocks or other shells in the intertidal zone.

Phylum Arthropoda
Class Crustacea
Order Throacica
Balanus spp. (Acorn Barnacles)

Table 11.2 Abundance of Shellfish Taxa by Unit and Layer (in grams/liter)

Unit	Layer	Volume Sampled (liters)	Clinocardium nuttallii (Heart Cockle)	Clinocardium spp. (Cockle)	Glycymeris subobsoleta (Pacific Coast Glycymeris)	Tresus spp. (Horse Clam)	Mya arenaria (Softshell Clam)	Mytilus spp.	Ostreidae (Oyster)	Chlamys rubida (Hind's Scallop)	Pecten caurinus (Giant Pacific Scallop)	Macoma nasuta (Bent-nosed Clam)	Unio margaritifera (Freshwater Clam)	Protothaca staminea (Littleneck Clam)	Saxidomus giganteus (Butter Clam)	Saxidomus spp.	Unidentified Venus Clam	Unidentified Bivalve	Littorinidae (Periwinkles)	Lottidae (Limpets)	Nassaridae (Nassas)	Polinices spp. (Moon Snail)	Thais spp. (Dogwinkle)	Unidentified Gastropod	Polyplacophora (Chitons)	Unidentified Mollusc	Balanus spp. (Barnacle)	Cancer spp. (Crab)	Strongylocentrotus spp. (Sea Urchin)	Total Shell Volume (g/l)
22/58	2A‡	10	-	*	-	-	-	-	-	-	-	-	-	-	-	-	0.1	0.2	-	-	-	-	-	*	-	0.1	*	-	-	0.4
	2B‡	36	0.1	-	-	-	-	-	-	-	-	-	-	-	-	-	0.2	0.1	-	-	-	-	-	-	-	*	-	-	-	0.4
	2C‡	16	*	*	-	*	-	-	-	-	-	-	-	-	-	-	0.6	0.3	-	-	-	-	-	*	-	-	*	0.1	-	1.2
	2D	12	0.4	0.4	-	0.3	-	0.1	-	-	-	-	-	0.4	-	0.2	4.0	2.2	-	-	-	-	-	-	-	3.3	0.5	-	-	11.8
	2E	14	0.2	0.6	-	0.2	-	0.3	-	0.1	-	-	-	1.2	0.4	-	4.3	7.3	-	-	-	0.2	0.2	*	-	5.3	0.8	-	-	21.2
	2F	6	0.7	0.2	-	0.3	-	0.3	-	-	-	-	-	1.2	-	-	6.0	11.9	-	-	-	-	-	-	-	5.3	0.9	-	-	27.0
	2G	8	0.2	0.1	-	0.1	-	0.1	-	-	-	*	-	0.1	-	-	0.9	4.0	-	-	-	-	-	-	-	1.4	0.1	-	-	6.9
	3A	8	0.3	0.2	-	-	-	-	-	-	-	-	-	-	0.3	1.1	1.0	1.1	-	-	-	-	-	-	-	1.0	0.1	-	*	4.7
	Unit Total		0.2	0.2	-	-	-	-	-	*	-	-	-	0.2	0.1	0.1	2.3	2.3	-	-	-	*	-	*	-	1.6	0.2	*	-	7.0
26/57	2A‡	22	*	*	-	-	-	-	-	*	-	-	-	-	0.1	*	0.4	0.6	-	-	-	-	-	-	-	0.3	*	-	-	1.5
	2B	10	-	0.1	-	0.4	-	-	-	-	-	-	-	0.1	-	-	0.3	0.9	-	-	-	-	-	-	-	0.8	0.1	-	-	2.7
	3A	4	*	-	-	-	-	-	-	-	-	-	-	*	-	-	0.8	0.5	-	-	-	-	-	-	-	0.2	-	-	-	1.5
	Unit Total		*	*	-	*	-	-	-	-	-	-	-	*	0.1	*	0.4	0.6	-	-	-	-	-	-	-	0.4	*	-	-	1.8
28/58	2A‡	22	-	0.3	-	0.1	-	*	*	-	-	-	-	0.2	-	0.1	3.2	1.0	*	*	-	0.1	0.1	*	-	3.2	0.4	*	-	8.6
	2B‡	24	0.3	0.4	-	0.4	-	*	*	-	-	-	-	1	0.2	0.2	20.1	4.7	*	-	-	0.9	0.1	0.1	-	11.6	1.6	-	-	40.6
	2C‡	32	0.5	0.2	-	-	-	*	-	-	-	*	-	1.2	0.1	0.1	20.2	8.9	*	*	-	-	-	0.1	-	10.6	1.1	-	-	43.7
	3A	16	0.4	0.3	-	1.8	-	0.3	*	-	-	-	-	1.2	0.1	-	3.7	6.4	*	*	*	0.3	0.3	*	*	5.7	0.5	-	*	20.5
	Unit Total		0.3	0.3	-	-	-	0.1	-	-	-	-	-	0.9	-	0.1	13.4	5.6	*	*	-	0.3	0.1	*	-	8.3	0.9	*	-	30.7
29/58	2A‡	20	0.2	0.5	-	*	0.1	*	-	-	-	-	-	0.5	-	-	2.2	2.7	*	*	-	-	-	*	-	5.2	1.3	-	-	12.9
	2B‡	12	0.1	0.2	-	-	-	0.1	-	-	-	-	-	0.1	-	-	2.1	3.8	-	-	-	-	-	-	-	5.1	1.0	-	-	12.6
	2C‡	24	1.9	2.5	-	1.2	-	0.2	*	-	-	-	-	0.8	-	1.0	9.7	9.0	*	-	-	1.0	0.1	*	-	14.1	0.6	-	-	40.4
	2D	32	0.9	3.0	-	2.0	-	0.7	-	-	-	-	-	6.8	-	1.5	39.5	20.0	-	*	-	-	-	*	-	21.6	2.6	-	-	99.6
	2E	38	1.3	2.2	0.2	1.9	-	1.5	-	-	0.2	-	-	9.9	0.5	0.6	30.9	8.7	*	-	-	0.1	-	0.1	-	11.5	2.8	-	-	75.6
	2F	24	0.6	2.0	-	5.7	-	0.3	*	-	-	*	0.5	4.4	0.1	1.6	15.1	10.7	*	*	*	0.1	-	0.1	*	9.6	1.1	-	-	51.8
	Unit Total		0.9	2.0	*	2.0	*	0.6	*	*	*	*	0.1	4.9	0.1	0.7	20.6	10.3	*	*	*	0.4	0.1	*	*	12.4	1.8	*	*	57.7
Total		394	0.5	0.9	*	0.9	*	0.3	*	*	*	*	0.1	2.1	0.1	0.7	11.7	6.0	*	*	*	0.2	*	*	*	7.2	1.0	*	*	31.4

‡ = layers thought to date to post-contact with Euro-Americans; all other layers date to pre-contact (see Table 5.1). * = < 0.1 liter.

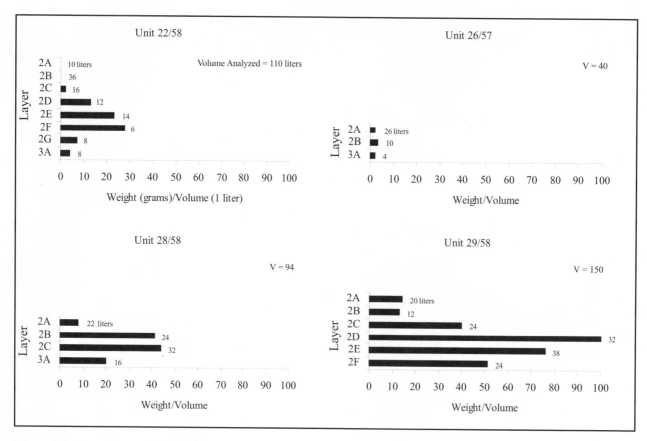

Figure 11.1 The amount of shellfish in each layer of each excavation unit was measured as weight and divided by the volume to standardize the comparison across layers.

Material: basal, opercular, and interlocking plates

Total: 386g

Remarks: Several barnacle species inhabit the rock or other hard substrates of the intertidal zone in Puget Sound. *Balanus cariosus* and *Balanus glandula* are both common to archaeological sites. They were not distinguished in this analysis. Nearly all the barnacles identified were too small to have been eaten, and were more likely to have been attached to other larger shellfish.

Class Malacostraca
Order Decapoda
Family Cancridae
Cancer spp. (Crabs)

Material: exoskeleton fragments

Total: 1.3g

Remarks: Few of these delicate fragments of crab were recovered from the site. Crabs occupy rock, gravel and kelp beds in the intertidal zones of bays and estuaries. They can be found as deep as 75 meters subtidally. No fragments were large enough to identify to species.

Phylum Echinodermata
Class Echinodea
Stronglyocentrotus spp. (Sea Urchins)

Material: exoskeleton fragments (test and spine)

Total: 0.1g

Remarks: Although sea urchin exoskeletal parts are easy to distinguish, few were found in the Burton Acres Shell Midden assemblage. The small number of sea urchins recovered from the site is likely a result of the lack of their preferred habitat nearby the site; they occupy rocky substrates in moderate to swift currents. Three species are common to the Northwest Coast: *Stronglyocentrotus franciscanus*, *Stronglyocentrotus droebachiensis*, and *Stronglyocentrotus purpuratus*.

RESULTS

Analysis of the shellfish at the Burton Acres Shell Midden provides an understanding of the way in which people at the site exploited the variety of taxa available in Quartermaster Harbor in the past 1000 years. A list of the taxa represented in the midden by weight per volume indicates the types of shellfish procured, as well as the relative abundance of these taxa (Table 11.2).

As much as 58% of the shell analyzed could be identified to family level or finer taxonomic level. Thirty-seven percent is identified to the Family Veneridae, the Venus clams. Within this family, nearly two thirds of the shells that could be identified to genus or species are *Protothaca staminea* (Native Littleneck Clam), and the remaining third are *Saxidomus* spp. (Butter Clam).

The other shellfish that could be identified to genus comprise about 12% of the shell within the site. Of these, the most common are (in order of abundance):

Clinocardium spp. and *C. nuttalli* (Cockle, Basket Cockle)
Balanus spp. (Acorn Barnacle)
Tresus spp. (Horse Clam)
Mytilus spp. (Mussel)
Polinices spp. (Moon Snail)

It is noteworthy that *Balanus* spp. is very easily identified due to its distinctive shell structure, and may be over-represented here. Nearly all of the *Balanus* spp. identified were whole or fragmentary pieces of very small specimens. These specimens are too small to have been procured for eating, and likely were incidentally collected because they were attached to larger shellfish species, such as clams and mussels.

A comparison of the taxa by weight/volume (Figure 11.1) between the four units and among the layers yields data that support a gradual change of species exploited through time from the prehistoric layers to the historic layers at the site. In Unit 22/58, the lowers layers (2D-3A) represent 97% of the shell found in that unit, and in Unit 29/58, 77% of the shell is found in those lower layers, although the decrease in shell abundance through time is a trend found throughout the excavated portion of the site. In Unit 28/58, a considerable amount of shell was found in Layers 2B and 2C, however, the amount of shell does decrease towards the uppermost portion of the unit.

Screen-Size Distribution

Figure 11.2 shows the distribution of shell remains by 1 inch, 1/2 inch, 1/4 inch and 1/8 inch mesh-sizes. In Units 28/58 and 29/58 the data indicate that more of the shell pieces are small in the upper layers (2A-2C), and more shell pieces are large in the lower layers (2D-2F). This size differential in the upper layers could be explained in several ways. One possibility is that the types of taxa procured in the upper layers are more fragile or smaller in size than in the lower layers. Another possibility is that certain activities performed in the harvesting or processing of the shell could have differentially fragmented the shell. Preservation could be a factor, given that the lower layers were inundated with saltwater at high tide. Or lastly, the difference could be a result of post-depositional activities such as cars driving over the site.

To some degree, Table 11.3 supports the first hypothesis of smaller-sized or more fragile taxa being in the upper layers.

Table 11.3 lists the percentages of taxa by layer for Units 28/58 and 29/58 (Units 22/58 and 26/57 are not included because of their small sample size). The data indicate an increase in both smaller taxa and more fragile taxa in the upper layers. In Unit 29/58, Layers 2A and 2B, are comprised of a larger percentage of barnacles (10% and 7.7%, respectively), when compared to the lower layers of Unit 29/58, which are comprised of less than 4% barnacle. The barnacles recovered at this site are very small (nearly all in the 1/4 inch and 1/8 inch), and increased amounts of such small barnacles would skew the distribution towards the smaller screen sizes. In terms of shell fragility, the data indicate that there is a larger percentage of robust shell in the lower layers. *Tresus* spp. (Horse clam) is represented in higher percentages in the lowest layer of both units (Unit 28/58, 3A and Unit 29/58, 2F). *Protothaca staminea* (Littleneck clam) is also more abundant in the lower layers (Unit 28/58, 3A and Unit 29/58, 2D, 2E, and 2F).

The second possibility, that of activity change, also appears to be a factor. The next section below, Shell Condition, indicates that more shellfish were burned in the upper layers than in the lowers layers.

Preservation as a result of high tide inundation is probably not a factor because the data indicate that smaller and more fragile taxa were found in the dryer, upper layers. If the lower layers are better preserved, one would expect fewer fragile taxa in the upper layers.

The fourth hypothesis, that of post-depositional activity, may also be a possibility at this site. Aerial

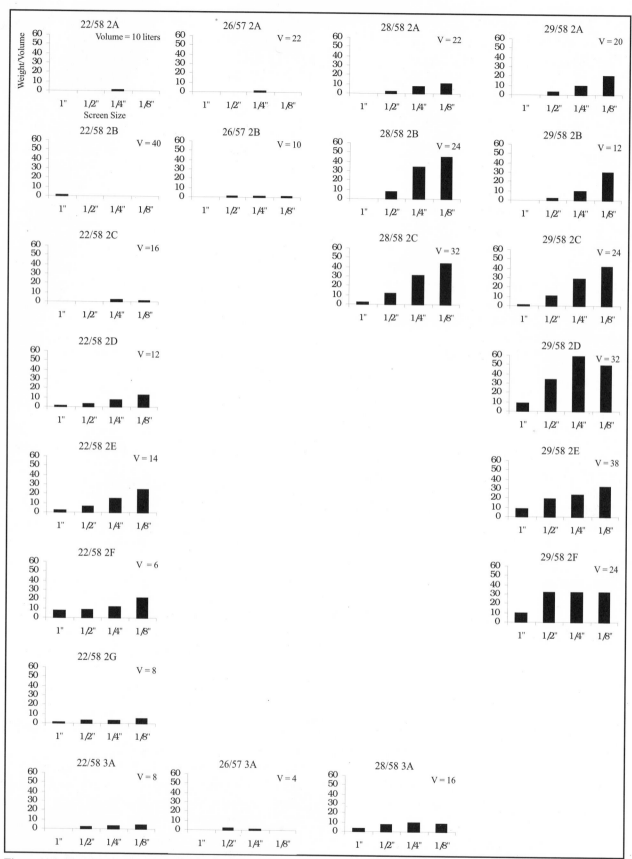

Figure 11.2 The weight of shellfish remains found in each screen size is displayed by layer and excavation unit. Bars represent weight (grams) per volume (number of excavated liters analyzed) to standardize the comparison across screen size.

Table 11.3 Percentage of Shellfish Taxa per Layer in Units 28/58 and 29/58 by Weight

Taxon	Unit 28/58				Unit 29/58					
	2A‡	2B‡	2C‡	3A	2A‡	2B‡	2C‡	2D	2E	2F
Mollusca (Total)	93.9	95.5	95.5	96.0	90.0	91.5	98.6	96.4	95.3	97.2
Unidentified Mollusc	37.1	28.6	24.2	28.0	40.2	40.7	34.9	21.7	15.2	18.5
Bivalve (Total)	56.8	66.9	71.3	68.0	49.8	50.8	63.7	74.7	80.1	78.7
Unidentified Bivalve	12.1	11.6	20.4	31.3	21.0	30.2	22.1	20.1	11.5	20.7
Tresus spp.	0.7	1.1	0.1	8.6	0.2	0.3	3.0	2.0	2.6	10.9
Mytilus spp.	0.1	-	0.1	1.3	0.2	1.1	0.6	0.7	2.0	0.6
Clinocardium spp.	3.2	1.5	1.6	3.0	5.2	2.0	10.9	3.9	4.7	5.0
Veneridae (Total)	40.7	52.6	49.1	23.8	22.8	17.2	27.1	48.0	59.3	39.7
Unidentified Venus clam	37.5	49.6	46.1	18.2	17.2	16.3	22.5	39.7	40.9	29.2
Protothaca staminea	1.7	2.6	2.8	5.6	4.0	0.9	2.1	6.8	13.1	8.5
Saxidomus spp.	1.5	0.4	0.2	-	1.6	-	2.5	1.5	5.3	2.0
Other	-	0.1	-	-	0.4	-	-	-	-	1.8
Gastropoda (Total)	1.9	0.6	2.1	1.7	-	0.7	-	1.0	1.2	0.5
Unidentified Gastropoda	0.5	0.2	0.1	0.1	-	-	-	-	0.1	0.1
Other	1.4	0.4	2.0	1.6	-	0.7	-	1.0	1.1	0.4
Arthropoda (Total)	4.3	3.8	2.4	2.3	10.0	7.7	1.4	2.6	3.6	2.1
Echinodermata (Total)	0.1	-	-	-	-	-	-	-	-	-
Total	100	100	100	100	100	100	100	100	100	100

‡ = layers thought to date to post-contact with Euro-Americans; all other layers date to pre-contact (see Table 5.1).

photos taken during the 1960s indicate that a driveway may have intersected a portion of the site. However, Units 28/58 and 29/58, which are close to the water's edge, were probably beyond the impact of cars.

Shell Condition

Less than 2% of the shells were burned (see Table 11.4). Burned specimens were identified visually in one of three ways: the shell was charred black, the shell was dark gray throughout, or the shell was white on the outside with a dark gray interior. Almost all of the burned shell was concentrated in the upper three layers (2A-2C) of Units 28/58 and 29/58. The heaviest concentrations were in Layer 2B, with 10.3% of the shell in Unit 28/58 and almost 30% (29.9%) in Unit 29/58. The majority of this burned shell that could be identified beyond phylum or class, was identified as Veneridae (Venus clams), and suggests that a change in activities occurred in the upper, historic layers. This is also supported by other cultural material found in these layers, such as the higher concentration of charcoal, burned fish and mammal bones, and melted glass.

Biological Source

With a few exceptions, all the shellfish recovered at the Burton Acres Shell Midden could have been harvested in the intertidal zone in front of the site or within Quartermaster Harbor. Quartermaster Harbor is characterized by calm, shallow waters with areas of gravelly, sandy, and silty substrates, and variable salinity levels.

Two species of scallop were found in the site, *Chlamys rubida* (Hind's scallop) and *Patinopecten caurinus* (Giant Pacific scallop), which were not likely residents of Quartermaster Harbor. *Chlamys rubida*, recovered from Unit 22/58, 2E, can be found in shallow waters, but is common in water as deep as 200 meters. Three modified fragments of *Patinopecten caurinus* also were found. This scallop prefers waters of ten meters to as deep as 200 meters, and live specimens are not common in the harbor near the site. The area just outside the harbor, at the south end of the island, is quite deep, and the empty scallop shells could have washed ashore and been picked up off the beach. Hilary Stewart (1973:157) suggests that, "this shell could only

Table 11.4 Distribution of Burned Shell Across Taxa by Percent of Total Weight of Shell

Unit	Layer	Mollusca					Arthropoda	Total
		Bivalvia				Unidentified Mollusc		
		Cardiidae	Veneridae		Unidentified Bivalve			
		Clinocardium spp.	*Prothaca staminea*	Unidentified Venus clam			*Balanus* spp.	
22/58	2A‡	-	-	-	-	-	-	-
	2B‡	-	-	-	-	-	-	-
	2C‡	-	-	-	-	-	-	-
	2D	-	-	-	-	-	-	-
	2E	-	-	-	-	-	-	-
	2F	-	-	-	-	-	0.1	0.1
	2G	-	-	-	-	-	-	-
	3A	-	-	-	-	1.9	-	1.9
Unit Total		-	-	-	-	0.0	0.0	0.1
26/57	2A‡	-	-	-	-	1.0	-	1.0
	2B	-	-	-	-	-	-	-
	3A	-	-	-	-	-	-	-
Unit Total		-	-	-	-	0.6	-	0.6
28/58	2A‡	-	-	1.8	-	1.5	-	3.3
	2B‡	-	0.3	4.9	0.8	4.3	-	10.3
	2C‡	-	-	0.3	0.3	0.4	-	1.0
	3A	-	-	-	-	-	-	-
Unit Total		-	0.1	1.9	0.4	1.8	-	4.2
29/58	2A‡	-	-	0.5	0.5	0.2	-	1.2
	2B‡	0.1	-	6.3	6.4	16.4	0.7	29.9
	2C‡	-	-	1.0	0.9	1.5	0.2	3.8
	2D	-	-	-	-	-	-	-
	2E	-	-	-	-	-	-	-
	2F	-	-	-	-	-	-	-
Unit Total		0.0	-	0.2	0.2	0.5	0.0	1.0
Total		0.0	0.0	0.6	0.3	0.7	0.0	1.7

‡ = layers thought to date to post-contact with Euro-Americans; all other layers date to pre-contact (see Table 5.1).

be collected after being washed ashore by storms, and being large and attractive, but not abundant or readily available to all villages, it became another trade item."

Both interior and exterior sides of these three fragments are ground (Figure 11.3). Stewart (1973:157,162) reports the use of Giant Pacific scallop for earrings and pendants.

A single freshwater clam, *Unio margaritifera*, was recovered from Unit 29/58, 2F. The closest freshwater creek, Judd Creek, is less than a mile away. This creek, which is one of the largest creeks on the island, does not currently yield freshwater clams. But, it certainly is big enough to have supported them in the past. It is possible that the pollution today has prohibited their growth in the creek.

Introduced Species

Mya arenaria, Soft-shelled clam, is not native to the western United States, and was introduced to the region in the 1890s. A single valve fragment with hinge was recovered from Unit 29/58, 2A. Its provenience less than three centimeters from the surface is not unexpected, as other artifacts from this portion of the site date to the late 1800s and early 1900s.

CONCLUSIONS

Comparison with other sites in Puget Sound

The diversity of shellfish taxa recovered from the Burton Acres Shell Midden is similar to other archaeological sites in Puget Sound. However, larger sites, such as Duwamish No. 1 (45KI23) (Campbell 1981; URS Corporation Seattle and BOAS, Inc. 1987; Jermann et al. 1977), Old Man House (45KP2) (Gaston and Jermann 1975) and the West Point Site Complex (45KI428 and 45KI429) (Larson and Lewarch 1995) contain larger percentages of Butter clams and mussels. It is noteworthy that mussels are often underrepresented in sites due to their fragility, however, this is not the case at Burton Acres Shell Midden. Site preservation, especially in the lower layers, was excellent, and often the periostracum (a covering which overlays the exterior of the shell) was found on the shells. A fragile freshwater shell fragment was recovered, and even crab shell remains. Additionally, no distinctive purple hue (indicative of mussels) was noted during the excavation or in the profiles. The absence of mussels may be attributed to the scarcity of rocky substrates in the area around Burton Acres Shell Midden.

Interestingly, most of the excavated (and reported) sites occupied in the last 1000 years in southern Puget Sound contain smaller percentages of Littleneck clam than at the Burton Acres Shell Midden (Campbell 1981; Ford 1995; Gaston and Jermann 1975; Jermann et al. 1977; Larson and Lewarch 1995; Larson 1996; URS Corporation Seattle and BOAS, Inc. 1987). However, sites further to the north in Puget Sound contain similar percentages of Littleneck clam. At the Manette Site (45KP9) in Bremerton (Jermann 1983), and Camano Island State Park (45IS95), Littleneck clams are the most abundant taxon, as is the case at the Burton Acres Shell Midden.

The shellfish diversity at the Burton Acres Shell Midden is fairly consistent throughout the timespan of the site. This is a characteristic shared with nearly all the analyzed shellfish assemblages in Puget Sound that derive from occupations in the last 1000 years. Consistency in shellfish diversity was reported at the West Point Site Complex (45KI428, 45KI429:Component 5),

Figure 11.3 These three fragments of scallops have had their interior and exterior edges ground. Scallops grow in deepwater habitats not found in Quartermaster Harbor, but found in areas nearby. Their presence in Burton Acres Shell Midden may indicate transport from nearby areas or trade. The grinding suggests they were being modified for some purpose, perhaps being made into earrings or pendants.

as well as at the Manette Site (45KP9), Camano Island State Park (45IS95), Allentown (45KI431) (Larson 1996), Whitelake (45KI438) (Larson 1996) and Old Man House (45KP2).

Procurement Activities

Nearly all the shellfish identified from the shell midden at Burton Acres Shell Midden could have been procured in Quartermaster Harbor throughout the year. Ethnographically, shellfish taxa such as Littleneck clams and Butter clams were most often eaten fresh, while cockles and Horse clams were more often harvested and dried in the spring and summer for winter consumption or trade (Smith 1940; Suttles 1951:30). There is little direct evidence at the Burton Acres Shell Midden to determine what types of shellfish processing occurred. The presence of pine bark in significant quantities (see Chapter 12) may indicate shellfish steaming. And, the recovery of two stake-like branches driven into the midden is suggestive of processing activities related to drying fish or shellfish.

Analysis of the fauna and flora from the Burton Acres Shell Midden provides valuable insights into how people used the site. Shellfishing and fishing were the main focus of activities. The site provides an excellent source for gathering these foods, as well as for processing them. Shellfish, especially Littleneck clams, were abundant and easy to collect. One can imagine people at the site eating fresh clams while they waited for their herring to dry.

REFERENCES

Abbott, R.T.
1974 *American Seashells*. Litton Educational
Publishing, New York.

Bryan, A.L.
1963 *An Archaeological Survey of Northern Puget
Sound*. Occasional Papers of the Idaho State
University Museum, No. 11, Pocatello, Idaho.

Campbell, S.K., (ed.)
1981 *The Duwamish No. 1 Site: A Lower Puget
Sound Shell Midden*. Research Report No. 1. Office
of Public Archaeology, Institute for Environmental
Studies, University of Washington, Seattle.

Cohen, A.N., and J.T. Carlton
1995 *Nonindigenous Aquatic Species in a United
States Estuary: A Case Study of the Biological
Invasions of the San Francisco Bay and Delta*.
United States Fish and Wildlife Service, Washing-
ton, D.C., and The National Sea Grant College
Program, Connecticut Sea Grant, NOAA Grant No.
NA36RG0467.

Duncan, M.A.
1977 *A Report of Archaeological Investigations
within Camano Island State Park*. Reconnaissance
Reports No. 14. Office of Public Archaeology,
Institute for Environmental Studies, University of
Washington, Seattle.

Ford, P.J.
1992 Interpreting the Grain Size Distributions of
Archaeological Shell. In *Deciphering a Shell
Midden*, edited by J.K. Stein, pp. 283-325. Aca-
demic Press, San Diego.
1995 Appendix 6: Invertebrate Fauna. In *The
Archaeology of West Point, Seattle, Washington:
4,000 Years of Hunter-Fisher-Gatherer Land Use in
Southern Puget Sound*, edited by L.L. Larson and
D.E. Lewarch, pp. 1-44. Larson Anthropological/
Archaeological Services, Seattle.

Gaston, J., and J.V. Jermann
1975 *Salvage Excavations at Old Man House (45-
KP-2), Kitsap County, Washington*. Reconnaissance
Reports No. 4. Office of Public Archaeology,
Institute for Environmental Studies, University of
Washington, Seattle.

Jermann, J.V.
1983 *Archaeological Investigations at the Manette
Site 45-KP-9, Bremerton, Washington*. Reconnais-
sance Reports No. 42. Office of Public Archaeology,
Institute for Environmental Studies, University of
Washington, Seattle.

Jermann, J.V., T.H. Lorenz, and R.S. Thomas
1977 *Continued Archaeological Testing at the
Duwamish No. 1 Site (45KI23)*. Reconnaissance
Reports No. 11. Office of Public Archaeology,
Institute for Environmental Studies, University of
Washington, Seattle.

Kozloff, E.N.
1996 *Marine Invertebrates of the Pacific Northwest*.
University of Washington Press, Seattle.

Larson, L.L., and D.E. Lewarch (editors)
1995 *The Archaeology of West Point, Seattle,
Washington: 4,000 Years of Hunter-Fisher-Gatherer
Land Use in Southern Puget Sound*. Larson Anthro-
pological/ Archaeological Services, Seattle.

Larson, L.L. (editor)
1996 *King County Department of Natural Resources
Water Pollution Control Division Alki Transfer/
CSO Project: Allentown Site (45KI431) and White
Lake Site (45KI438 and 45KI438A) Data Recovery*.
Submitted to HDR Engineering, Inc., Bellevue,
Washington. Prepared by Larson Anthropological/
Archaeological Services.

Smith, H.I.
1907 Archaeology of the Gulf of Georgia and Puget
Sound. In *The Jesup North Pacific Expedition*, vol.
2, pt. 6, edited by Franz Boas. E.J. Leiden Ltd.,
New York.

Smith, M. W.
1940 *The Puyallup-Nisqually*. Columbia University
Contributions to Anthropology Vol. 32. Columbia
University Press, New York.

Stewart, H.
1973 *Indian Artifacts of the Northwest Coast*.
University of Washington Press. Seattle.

Suttles, W. P.
1951 Economic Life of the Coast Salish of Haro and
Rosaria Straits. Unpublished Ph.D. dissertation,
Department of Anthropology, University of Washington, Seattle.

URS Corporation Seattle and BOAS, Inc.
1987 *The Duwamish No. 1 Site 1986 Data Recovery*.
Submitted to: Municipality of Metropolitan Seattle
(METRO), Contract No. CW/F2-82, Task 48.08.
Prepared by BOAS, Inc., Seattle.

Wessen, G.C.

1994a Part III: Subsistence Patterns as Reflected by Invertebrate Remains Recovered at the Ozette Site. In *Ozette Archaeological Project Research Reports, Volume II, Fauna*, edited by Stephan R. Samuels, pp. 95-196. Reports of Investigations No. 66. Department of Anthropology, Washington State University, Pullman, and National Park Service, Pacific Northwest Regional Office, Seattle.

1994b An Account of the Ozette Shellfish Taxa. In *Ozette Archaeological Project Research Reports, Volume II, Fauna*, edited by Stephan R. Samuels, pp. 333-358. Reports of Investigations No. 66. Department of Anthropology, Washington State University, Pullman, and National Park Service, Pacific Northwest Regional Office, Seattle.

12

Botanical Analysis

Nancy A. Stenholm

Plant remains provide clues to peoples' food, fuel, and technology. Unfortunately, plants decay quickly when buried in the Northwest ground. But when charred, plants preserve well. At Burton Acres Shell Midden charred plants were found throughout the site. The charcoal of conifers (pine and Douglas fir) is abundant and indicates that the wood and bark of these trees were useful as fuel. Such quantities of charcoal, along with burned shellfish, fish and mammal bone, support the hypothesis that people came here to harvest and prepare food.

The study of vegetable materials found in archaeological matrices, termed archaeobotany or paleoethnobotany, provides valuable information about people inhabiting a site. Botanical material, along with lithic and faunal data, gives archaeologists the ability to interpret subsistence, site features, and season of use. The importance of botanical analysis in hunter-gather studies cannot be overstated. The field is new and growing rapidly. People who do not practice agriculture depend heavily on plants for nourishment. To determine the kinds of plants and their uses at this site, we used the special techniques described here.

METHODS

Although vegetable materials utilized by prehistoric people decay rapidly, evidence of plant gathering, preparation, and use is often preserved as charred, microscopically identifiable particles contained in soil samples. Because these small and friable materials are seldom recovered in situ or during routine screening, general processing procedures for botanical samples are different from those used for faunal material and lithics. Processing soil samples to extract botanical remains consists of six steps: (1) determining the weight and volume of the dried matrix, (2) separating using water (flotation) the light from the heavy fraction, (3) drying the two fractions, (4) further separation using chemicals, (5) drying the two fractions, and (6) weighing the light and heavy fractions.

Water separation consists of submerging a container of excavated material in water, allowing the clays, silts, and sands to wash out the bottom of a screen, and recovering light materials that float to the surface. Floating material (the light fraction) is removed from the surface of the water and both it and the residue at the bottom of the container (the heavy fraction) are dried. These two fractions comprise a

Chapter opening photo: At the sorting area volunteer Matt Chernicoff (right) separates charcoal from shell, bone, stone, and metal. He and his brother, David, are working together on a bucket that had to be sorted completely. The charcoal saved here was used for radiocarbon dating. However, in order to study the plants used by the occupants of the site more charcoal had to be collected in an entirely different way – through water separation, called "flotation."

flotation sample, or float sample.

Since some charcoal will not float without additives to increase solution density, the heavy fraction is submerged in liquid with a specific gravity of 1.2. The light fractions of both water and chemical flotations are combined before they are weighed. Subsamples of the combined light fraction are then extracted to get a sample of the appropriate size for detailed analysis. The light fraction is passed through nested screens and 0.25g of material caught between 5.0mm and 3.0mm mesh is completely analyzed by identifying all botanical remains found. Experience shows that for Northwest Coast shell middens, most soil samples weighing from one to two kilograms (approximately 1 to 2 liters) contain adequate amounts of charcoal (or other plant material) to allow interpretations. If a sample has a below average amount of charcoal, all available charcoal is picked by hand from the sample and weighed.

Identification is done with the aid of a binocular microscope with continuous magnification from 10 to 150 times. Weights are taken in milligrams rounded to the nearest hundredth of a gram. A primary distinction is made between charred (or semi-charred) material and non-charred material since the carbonized material resists decay for considerable periods of time. The non-charred material is more likely to represent recent (within months or, at most, years) contributions incorporated into the sample matrix through bioturbation (mixing of soil by organisms).

Microscopic analysis begins with manual separation and weighing of charred and semi-charred botanical materials from all other materials in the subsample. These weights are used to estimate the total carbon or the total non-charred flora in the sample (the "carbon/non-charred flora percentage or carbon/flora content"). Both figures are useful in predicting the presence of certain kinds of features associated with burning, such as occupation floors, hearths, and trash pits. Mixing by plants and animals may also be indicated by the presence of insect parts, modern flora, or rodent remains.

Analysis proceeds by dividing the carbonized material into woody material, seeds, surface and subsurface fruit, root, and stem tissue, and dissociated tissue types. For the most part, only family and genus identifications can be made on wood, seeds, bark, and certain portions of stem tissue (including bud, flower, leaf, and fruit fragments). For example, conifer needles preserve well and are

relatively easy to identify compared with the more fragile leaves of hardwood species. Fern stems are easily identified compared with the softer stems of most herbaceous plants. Tissues rarely can be identified taxonomically, but the distinction of general tissue types can be important in assessing preservation factors as well as presence of processing and technological activities. Presence of fibers and bark tissues, for instance, may indicate cordage production. Seed coat fragments may indicate fruit or seed processing associated with grinding or pounding. Other tissues may be remnants from processing soft parts of foods and medicines.

Tissues form a special category because they can be divided into groups based on certain characteristics. A glassy or shiny material may represent plant saps, juices, or resins. It is amorphous black or dark brown material with bubble or steam cavities. When it is found with wood cells it is most likely wood pitch or conifer resin. When it is associated with other softer tissue types, such as starchy parenchymoid or fruity epithelioid tissues, it may be processed edible tissue (PET). In this analysis, the PET category is divided into three groups: tissue that resembles sugar-laden fruit or berry tissue with seeds (PET fruity), fruit or berry tissue without the seeds (PET other), and tissue with starchy storage cells (PET starchy) probably from edible roots.

Analysis of a flotation sample is completed with a scanning of the remaining (non-subsampled) portions of the light and heavy fraction for diagnostic pieces of charcoal, lithics, bone, shell, and other cultural material. Diagnostic floral material is recorded as a trace (less than 0.01g). Lithics, bone, and shell are weighed and are counted. The presence of burned earth, pigment, and historic materials is noted. Modern rodent, insect and plant remains are saved to determine the population active in the soil environment. Ancient bioturbation is interpreted when charred insect remains are found.

RESULTS

This report concerns seven flotation samples from Unit 29/58 at Site 45KI437, Burton Acres Shell Midden on Vashon Island. Table 12.1 lists the flotation samples including weight, volume, carbon found in grams, and the weights of shell, bone, iron, glass, and flakes for each sample. At least 14 plant taxa were found in 129g of archaeobotanical materials extracted from 12.3kg (9.0 liters) of site matrix. These values are comparable to other

Table 12.1 Sample Summary, Unit 29/58

Unit Level (cm below surface)	Weight (g)	Carbon		Shell	Bone	Iron	Glass	Flakes	
		g	C%	g	g	g	g	#	g
3-5	2055	8.3	0.4	156.3	0.2	1.2	12.3	1	0.1
10-15	1365	43.3	3.2	44.7	7.6*	9.2	0.3	-	-
20-25	1570	27.2	0.2	82.6	1.5*	23.2	0.1	-	-
30-35	1790	16.4	0.9	191.1	6.9	trace	-	-	-
40-45	1865	2.2	0.1	308.9	4.4	-	-	3	2.3
50-55	2130	18.2	<0.1	224.9	11.5	-	-	1	1.0
60-65	1530	14.4	0.9	7.7	0.2	-	-	-	-
Total	12310	129	1	1016.2	32.3	33.6	12.7	5	3.4

* = Eggshell present.

sites in the region (Table 12.2). Archaeobotanical flotation results are shown in Table 12.3.

Shell is the most abundant material type by weight in all but one sample (60-65cm), followed by charred archaeobotanical remains (carbons). All samples contained shell fragments of clams, mussel, barnacle, and some oyster. One flotation sample required eight hours to remove all fragments of shell and bone larger than the 0.5mm screen size. Five samples had so many tiny fragments of bone and shell that subsamples were drawn in order to estimate the total amount of bone and shell smaller than the 4.75mm screen size. The bone and shell from levels 10cm to 15cm below surface and 60cm to 65cm below surface are actual weights of material removed by hand. All samples contain bone (at least two kinds of fish and some mammal bone). Approximately half have historic artifacts such as rusted iron and glass. Three samples contain lithic flakes. A fragment of disintegrating rubberized fabric tape was identified from one flotation sample, and looks to be a few years old. Bone, flakes, fire-modified rock (FMR), burned earth, glass and iron were also found.

All flotation samples have charred and non-charred insect remains and all contain modern floral material including wood, needles, bark, roots, and rootlets in all stages of decay. The samples taken from the deepest and shallowest locations contain polyester fibers, probably from excavation activity. From these data, bioturbation is interpreted to be extensive (heavy) in the deposits from the surface to 35cm below surface, and moderate to light in the deposits from below 35cm to 65cm below surface.

Conifer

Conifer charcoal consisting of wood and bark constitutes 64% of the assemblage weight from the shell midden. Each flotation sample contains identifiable taxa. Red cedar (*Thuja plicata*), hemlock (*Tsuga heterophylla*), a cedar or hemlock taxa, and an unidentified conifer are represented by wood, and the Pine family (Pinaceae) and Douglas fir (*Pseudotsuga menziesii*) are represented by bark. Both wood and bark are found in every sample. In addition to charcoal, traces of Douglas fir needles and cone brack were found in three samples.

Identifiable pieces of red cedar and hemlock are found in one sample each. Segments of branches, however, of these two taxa are very difficult to separate accurately and are listed here as cedar or hemlock (cedar/hemlock). Branches have been identified from three samples. A small, worn, charred fragment of conifer wood was found in the sample from the depth 60cm to 65cm below surface. It is approximately 4.5mm long, 1mm wide, and 0.5mm thick (at 0.01g). It may have been rolled before or after charring by natural beach action (making identification problematic) and probably is not associated with human activity.

Red cedar was used extensively by Native peoples for a variety of purposes (Gunther 1945:19-21; Turner 1979:74-90; Turner et al. 1983:67-70). Red cedar was used extensively, while hemlock apparently was not considered good fuel, especially when green. Hemlock could be substituted in place of yew in halibut fishing apparatus, and knots from logs were made into halibut hooks (Turner et al. 1983). Branches were used in collecting herring spawn. The oldest Douglas fir wood and

Table 12.2 Plant Taxa Yield From Selected Sites, Puget Sound and Pacific Coastal Regions, Western Washington

Site		Approximate Dates (BP)	Material Floated (kg)	% Carbon	Number of Taxa	Taxa/kg
Number	Name					
45KI437	Burton Acres	1000-100	12	1.0	14	1.2
45KI435	Mule Spring	5000-100	5	2.7	10	2.0
45KI23	Duwamish No. 1	2000-1000	80	0.4	41	0.5
45KI431	Allentown	600-100	6	0.3	16	2.6
45KI438/438A	White Lake	2000-200	3	0.5	12	4.0
45KI428	West Point	4000-200	61	0.6	20	0.3
45KI429	West Point	4000-200	49	1.5	24	0.5
45KI125	Chester Morse Lake	8500	12	0.7	10	0.8
		2000-1000	12	1.0	20	0.8
45LE226	Burton Creek Rock Shelter	1000	4	1.5	15	3.8
45LE223	Layser Cave	6600-6400	14	0.2	8	0.6

Adapted from Stenholm (1987; 1989a, b; 1990; 1993; 1995a, b; 1996).

bark found in Puget Sound is from a sample collected within a rock feature at 45KI125 with a radiocarbon date of 8500 BP (Stenholm 1993:Table 12.3).

Bark (as opposed to wood) is 34% of the assemblage weight. Pine family (Pinaceae*)* bark is the most abundant plant material by weight, followed by Douglas fir. Much of the Douglas fir bark is probably from mature trees.

Pine family bark is prominent at the West Point sites. It is particularly abundant at Site 45KI428 and contributes 48% of the assemblage weight (Stenholm 1995a:Appendix 7, Table A7.3) and 28% at 45KI429 (Stenholm 1993:Table A7.8). The bark was used primarily as fuel and has been linked with steaming and drying shellfish and fish (Larson 1995:13-27). The Burton Acres Shell Midden resembles the West Point Site being the second time large amounts of bark and shellfish have been found together. Douglas fir bark is considered a good all-around fuel (Turner et al. 1983:22,73).

Hardwood

Hardwoods comprise 27% of the assemblage weight. The most abundant taxon is hardwood bark, followed by oceanspray (ironwood, *Holodiscus discolor*), willow (*Salix* sp.), traces of maple (*Acer circinatum* or *A. rubrum*), and poplar (*Populus* sp.) or willow. Also found are traces of a diffuse-porous hardwood that cannot be identified further, but is mostly likely madrone (*Arbutus menziesii*).

All hardwoods found are used by Native American peoples in the Northwest and all (except possible madrone) have been encountered in other archaeological analyses. Oceanspray, a bushy species, is ubiquitous in

Western Washington archaeological sites. As a bushy species, it is not a likely candidate for fuel. The spring wood is useful, however, in the manufacture of shafts, prongs, foreshafts, clam-drying sticks, hooks, digging sticks for clams, and other items (Gunther 1945:33; Turner 1979:234-236; Turner et al. 1983:117-118). The wood is good for barbecuing because it does not impart unpleasant flavors to the food. The oldest oceanspray found in flotation samples is from a 6000-6500 year old hearth in Layser Cave, Washington (45LE223) (Stenholm 1989b).

The tough flexible branches and stems of vine maple and Rocky Mountain maple (*Acer circinatum* and *A. labrum* respectively) are made into large open-work carrying baskets, fish traps, bows, arrows, frames and implement handles, as well as prongs, pins, and the like (Gunther 1945:40; Turner 1979:154-159). Maple is the most abundant hardwood at the Duwamish No. 1 Site (45KI23) (Stenholm 1987:13-19). At the Chester Morse Sites it is the third most abundant after poplar and willow (Stenholm 1993).

Willow is a utilitarian species useful as medicines and in flexible constructions such as hoops, traps, baskets, and cordage (Turner 1979:260-265). Some willow is nearly always found in archaeological assemblages. Willow and poplar often cannot be assigned to genus (branch material mostly) and it is found in a few samples. As with similar material in the cedar or hemlock categories, the term Poplar/Willow (pop/wil) is used in this study. It is found in one sample.

Table 12.3 *Botanical Assemblage of Unit 29/58 by Flotation Weight (g) and Number of Appearances (#).*

Botanical Remains	Level (cm below suface)														Total	
	3-5		10-15		20-25		30-35		40-45		50-55		60-65			
	g	#	g	#	g	#	g	#	g	#	g	#	g	#	g	#
Conifer (64%)																
Red cedar	-	-	-	-	-	-	-	-	-	-	0.02	1	-	-	0.02	1
Hemlock	-	-	-	-	0.02	1	-	-	-	-	-	-	-	-	0.02	1
Hemlock/cedar	-	-	0.01	1	0.02	1	0.01	1	-	-	-	-	-	-	0.04	3
Douglas fir	<0.01	1	0.01	1	0.02	1	0.03	1	<0.01	1	0.03	1	0.02	1	0.11	7
Bark	-	-	0.02	1	0.03	1	0.05	1	0.05	1	0.12	1	-	-	0.27	5
Needles	-	-	<0.01	1	<0.01	1	<0.01	1	-	-	-	-	-	-	<0.01	3
Cone	-	-	0.01	1	-	-	<0.01	1	-	-	-	-	-	-	0.01	2
Twigs	-	-	-	-	<0.01	1	-	-	-	-	-	-	<0.01	1	<0.01	2
Hardwood (27%)																
Maple	-	-	-	-	-	-	<0.01	1	-	-	-	-	-	-	<0.01	1
Oceanspray	-	-	<0.01	1	-	-	0.01	1	-	-	-	-	-	-	0.01	2
Willow	0.01	1	-	-	-	-	-	-	-	-	-	-	-	-	0.01	1
Poplar/willow	-	-	<0.01	1	-	-	-	-	-	-	-	-	-	-	<0.01	1
Diffuse-porous	<0.01	1	-	-	-	-	-	-	-	-	<0.01	1	-	-	<0.01	2
Bark	-	-	-	-	0.01	1	0.01	1	-	-	-	-	0.16	1	0.18	3
Edible Tissue (<1%)																
Elderberry, 2	-	-	-	-	-	-	-	-	-	-	-	-	<0.01	1	<0.01	1
Rubus, 1	-	-	-	-	-	-	-	-	-	-	-	-	<0.01	1	<0.01	1
PET fruity	-	-	<0.01	1	-	-	-	-	-	-	-	-	-	-	<0.01	1
PET starchy	-	-	<0.01	1	-	-	-	-	-	-	-	-	-	-	<0.01	1
Other Tissue (8%)																
Seeds, 10	-	-	<0.01	1	-	-	<0.01	1	<0.01	1	-	-	<0.01	1	<0.01	4
Herb stem	-	-	0.01	1	-	-	<0.01	1	-	-	-	-	0.03	1	0.04	3
Monocot	-	-	-	-	-	-	-	-	-	-	-	-	0.03	1	0.04	2
Root	-	-	<0.01	1	<0.01	1	<0.01	1	-	-	-	-	-	-	<0.01	2
Paryenchymoid	-	-	0.01	1	-	-	-	-	-	-	-	-	-	-	0.01	1
Other	-	-	0.01	1	-	-	-	-	-	-	-	-	-	-	0.01	3
Total	0.01		0.08		0.10		0.11		0.05		0.17		0.21		0.73	

Note: One flotation was performed per level.

Table 12.4 Botanical Arrays of Sites Depicted by Floral Category

Site		Floral Category by Weight (% of Sample)			
Number	Name	Conifers	Hardwoods	Edible Tissue	Other Material
45KI437	Burton Acres	64	27	<1	8
45KI438/438A	White Lake	14	49	14	24
45KI431	Allentown	63	17	17	3
45KI23	Duwamish No. 1	67	11	12*	10*
45KI428	West Point	69	20	3	8
45KI429	West Point	71	19	2	8
45PC101	North Nemah Bridge	82	9	6	6
45KI125	Chester Morse Lake	77	14	<1	7
45LE266	Burton Creek Rockshelter	86	14	1	5
45LE223	Layser Cave	37	25	2	36

*These are not the original report figures (which were 51%, 11%, 2%, and 36%). In the 1987 report, PET was not included in the Edible Tissue category (it was hinted that it might be placed there in future). Duwamish edible category should be increased to 12% and decrease the Other Tissue category to 26%. Additionally, a tissue type (which weighed 1.71g) has been added to the conifer bark category, which raises the Conifer total and lowers the Other Tissue category further.

Finally, hardwoods that cannot be identified to family appear as diffuse-porous specimens and are present in two samples. Some is likely to be madrone (*Arbutus menziesii*). This identification is tentative and is the first of its kind; more charred wood needs to be examined.

Edible Tissue

Specimens of the edible category of tissue are not frequently found at Burton Acres Shell Midden. The ones found are two elderberry seeds (*Sambucus* sp.), a partial raspberry (*Rubus* sp.) seed, and processed edible tissue (PET) of both fruity and starchy tissue.

The two elderberry seeds are found in the sample from 60cm to 65cm below surface level. They may be as old as 1000 years. Elderberry fruit seeds are commonplace in flotation samples from Oregon and Washington, but are not necessarily products of human collection or consumption. Charred and non-charred elderberry seeds are found in many coastal locations, and elderberry bushes are probably associated with disturbed soils found in habitation sites much like goosefoot (*Chenopodium* sp.) and other weedy plants that thrive in the enriched soils. Elderberries have been an important food resource in the Pacific Northwest for a long time (Gunther 1945:47; Turner 1975:125-127), and elderberries can be harvested from late June throughout the summer months depending on location and elevation. Turner notes that elderberries were once gathered in late July and August, cooked and stored as dried cakes to be served later with other foods (Turner 1975:125-126). Elderberries are good seasonal

indicators and suggest that people collected them during the summer. Because the elderberries are found with other abundant indicators of human activity, they point strongly to being collected from nearby locations. Currently the oldest elderberry seeds and fruit tissues found in Puget Sound are from the West Point Site in a layer dated to 4250-3550 BP (Stenholm 1995a:7-16).

A small, partial *Rubus* seed (blackberry or wild raspberry) was found with the elderberry seeds. The achene is 1.5mm long, 1.1mm high, and 1.0mm wide. The surface is eroded and species cannot be determined. Rubus seeds are also common in western Washington sites but so fragmented that species identification is impossible. This sample is too small to be the Old World species (Himalaya blackberry, *R. procerus*) currently growing at the site. Blackberries ripen from June to July depending on exposure and elevation, and several native species including wild raspberry (*R. idaeus*), black raspberry or blackcap (*R. leucodermis*), thimbleberry (*R. parviflorus*), trailing raspberry (*R. pedatus*), salmonberry (*R. sectabilis*), and trailing wild blackberry (*R. ursinus*) were sought as food by coastal groups (Turner 1975:212-223; Turner et al. 1983:123-125).

Lastly, there are two specimens of PET tissue from a flotation sample taken at the depth of 10cm to 15cm below surface. One is fruity tissue with no seeds present. It is found with a large amount of amorphous "glop" described below as amorphous glassy material. The second is PET starch and may be the starch from nutmeat rather than the less dense starch of storage roots. In Western Washington and Oregon, nutmeat is rarely found in archaeological sites. Nuts are usually represented by hulls (e.g. hazelnut [*Corylus cornuta*]) or shells (e.g. White oak acorn [*Quercus garryana*]), which are not found at this site.

Other Tissue

Four flotation samples contained seeds that could not be identified as edible tissue (ten seeds representing four taxa) including three small chenopodium seeds, five small oval seeds (mostly 1-1.5mm in diameter) probably from the legume family (Apiaceae, Leguminosae), and two seed coat fragments from two other taxa that cannot be further identified.

Four flotation samples contain herbaceous stem tissue, two have small roots, and one has parenchymoid tissue with large cells suggesting plant pith (e.g., from the center of woody plant branch).

Finally, three samples have amorphous glassy material. The 10cm to 15cm sample has bubble casts and steam-cavities. There is 13% by weight and it is found in association with PET fruity tissue. The material may be hardened sugary sap or other kinds of drippings (for a discussion, see Stenholm 1987).

DISCUSSION

The flotation samples provide results that can be summarized through time. Each sample is arranged in chronological order with the shallowest being the most recent and the deepest the oldest. The plant data suggest certain subsistence activities took place at the Burton Acres Shell Midden. The notation "Nas" is the abbreviation for "Nancy A. Stenholm number."

Level 3cm to 5cm below surface (Nas#9)

The flotation sample from the uppermost level has an estimated 8.32g of archaeobotanical materials and a carbon content of 0.4%. There are three plant taxa present (for a yield of 1.5 taxa/kg). The sample is approximately 99% willow and less than 1% each of Douglas fir and diffuse-porous hardwood (possibly madrone, *Arbutus menziesii*).

The sample contained mammal and fish bone, a moderate amount of shell, burned earth, and rusted iron fragments. There is glass of several colors (clear, yellow, amber and green) and a possible pinkish flake. The sample also had a small fragment of white rubberized tape, 11.5mm wide, with pinked edges and holes with fragments of thread adherent.

Bioturbation is moderately severe with roots, conifer wood, grass, bark, modern seeds, and insect fragments, polyester fibers and modern Old World blackberry (Himalaya blackberry, *Rubus procerus*) seeds.

Activities interpreted from this sample are the present uses as a picnic area mixed with the historic use as a shellfishing site. The activities continue today.

Level 10cm to 15cm below surface (Nas#6)

The sample from Level 10cm to 15cm has the highest carbon content at the site at 3.2% and the greatest number of plant taxa (9 taxa/kg). The sample is 63% conifer, 37% other tissue, with traces of hardwood (oceanspray and poplar/willow) and edible tissues.

The most abundant conifer material is from the pine family (Pinaceae) bark, followed by Douglas fir, hemlock/cedar, a Douglas fir cone bract and Douglas fir fragments.

Edible material is represented by traces of PET fruity and PET starchy material.

The Other Category contains herbaceous stem tissue, rootlet tissue, two seed coat fragments (two taxa), paryenchymoid tissue that is likely woody plant pith and amorphous material with bubble cavities that cannot be identified further.

The sample has moderate amounts of porous or cancellous mammal bone and fish bone, glass, burned earth, and iron fragments (including a nail). It is one of two samples with eggshell (0.05g), and it has the second lowest amount of shell in the series.

The sample contained some unburned modern grass, conifer bark, modern blackberry, elderberry, and chenopod seeds. Bioturbation is moderate.

The botanical evidence suggests activities associated with shellfish utilization, including bark and wood used as fuel (Douglas fir), as well as woods with known subsistence uses. The PET material (possibly the amorphous material with bubble cavities), the two seeds coats and the eggshell point toward spring or summer occupancy.

Level 20cm to 25cm below surface (Nas#7)

This sample has a carbon content of 0.2%, and four plant taxa (2.5 taxa/kg). Conifer (mostly pine family bark) makes up 90% of the sample and 10% hardwood bark with traces of herbaceous stem tissue. The conifer category also contains hemlock, hemlock/cedar, conifer twigs, as well as Douglas fir and hemlock needles.

The sample has the largest number of rusted iron fragments, as well as moderate amounts of shell, mammal and fish bone, and some clear glass and burned sediment.

Eggshell is present.

Bioturbation is moderate and represented by confer bark and needles, grass, and modern seeds (chenopod and elderberry).

Level 30cm to 35cm below surface (Nas#8)

The flotation sample from the 30cm to 35cm level has a carbon content of nearly 1% and seven taxa (2.6 taxa/kg). The botanical remains are 82% conifer (mostly pine family bark) with Douglas fir, red cedar, and some hemlock/cedar branch tissue. A Douglas fir cone bract and traces of Douglas fir and hemlock needles are present. The sample is 18% hardwood including oceanspray, hardwood bark, maple, and twigs, and contains three chenopodium seeds (1.5mm in diameter, with pericarp), two small (1.1mm to 1.5mm across) oval seeds that cannot be further identified, and a trace of glassy tissue, which may be plant sap or resin.

The sample has an increase in shell and bone (fish and mammal) relative to the sample immediately above it. It also contains a trace of rusted iron (wire). Both FMR and burned earth were encountered.

Bioturbation is light in this sample and it includes modern and ancient insect dissemules.

Level 40cm to 45cm below surface (Nas#10)

The amount of floral material in this sample decreases substantially over the sample above it. The carbon content drops to 0.1% and the plant taxa to 0.9 taxa/kg. The plant remains are approximately 99% pine family bark with traces of Douglas fir and a small seed, which cannot be identified further.

Shell weight increases further and it is the highest of the series. Mammal and fish bone decrease slightly over shallower levels. FMR and burned earth are present, as well as, three possible lithic flakes (two are basalt).

Bioturbation remains light.

The sample probably represents an episode of shellfish processing (perhaps a discard pile). The light floral scattering indicates the sample was from the periphery of main heating/burning activity rather than part of the hearth itself.

Level 50cm to 55cm below surface (Nas#11)

This sample has a carbon content of 0.09%, and plant remains from three plant taxa (1.4 taxa/kg). The remains are 99% conifer (mostly pine family bark) and less than 1% hardwood. The conifer is Douglas fir and red cedar.

The sample has the highest abundance of bone (mostly fish) and second highest of shell. The sample

contained a black and clear-banded flake.

Bioturbation is not indicated in this sample.

The sample probably represents a prehistoric episode of shellfish and fish processing similar to the sample above.

Level 60cm to 65cm below surface (Nas#5)

There is an increase in archaeobotanical materials over the immediately shallower level. This sample has a carbon content of 0.9% and six taxa (4.0 taxa/kg).

The plants are 76% hardwood bark, 10% Douglas fir charcoal, less than 1% edible tissue, and 14% other tissue. A partial rubus seed and two elderberry seeds represent edible tissue. There is a large amount of herbaceous stem material present. Some appears flattened or slightly twisted before it was charred. This suggests that the material was dry and undergoing cellular collapse and that the stems were partially dry when burned. Finally, there are two small oval seeds that cannot be further identified.

Fish bone and shell is not abundant in this sample. No mammal bone is identified from the flotation.

Bioturbation is moderate. Evidence includes roots, conifer needles, rubus (seed), legume seed, mammal hair, polyester fibers, and paper bits. Most appears to be of modern origin.

The sample is different from most samples in this series. Hardwood bark (hardwood charcoal is lacking) has replaced conifer bark. Seeds from known edible fruits have made their appearance. And bone and shell content is much reduced. The sample is similar to samples from campsites (as opposed to processing sites) where several kinds of procurement activities took place. This contrasts with the samples above that have only a few specialized activities represented.

CONCLUSIONS

Some generalizations can be made concerning the Burton Acres Shell Midden archaeobotanical material. Archaeological preservation is excellent and the diversity of taxa is high. There is at least one major fuel present (Douglas fir wood and possible bark). Two conifers (cedar and hemlock) and three hardwoods that have well-documented industrial uses are also present (oceanspray, maple, and willow). On the average, one plant taxon is found for every kilogram of soil analyzed. The elderberry and rubus seeds indicate occupation in summer. The remnants of edible seeds, PET tissue, and the glassy material (with steam cavities) suggest plant gathering but not much processing. In general, the large amounts of shell associated with the equally large quantities of bark suggest that abundant shellfish and finfish processing took place on this shoreline.

The Burton Acres Shell Midden botanical assemblage resembles those of seasonally occupied sites such as the West Point sites (fishing and shellfish gathering), the Chester Morse Lake Site (upland seasonal hunting), North Nemah Bridge (a coastal fishing station with a probable domestic structure), Burton Creek Rock Shelter (a transit site for upland hunting and/or gathering) (Stenholm 1989a), and Layser Cave (and upland staging site for hunting and possible fruit gathering) (Stenholm 1990). In these samples the floral Edible Category is modestly represented, while at White Lake (Stenholm 1996), Allentown, and Duwamish No. 1 edible tissue weights are very high. The botanical remains found at Burton Acres Shell Midden represent plant gathering and consumption over several seasons.

REFERENCES

Gunther, E.

1945   Ethnobotany of Western Washington. *University of Washington Publications in Anthropology* Vol. 10. The University of Washington Press, Seattle.

Larson, L.L.

1995   Subsistence Organization. In *The Archaeology of West Point* vol. 1, pt. 2, edited by L.L. Larson and D.E. Lewarch, pp. 13.1-13.46. Larson Anthropological/Archaeological Services, Seattle.

Stenholm, N.

1987   Botanical Analysis. In *The Duwamish No. 1 Site, 1986 Data Recovery*, edited by URS Corporation and BOAS, Inc., pp. 13.1-13.31. Submitted to Municipality of Metropolitan Seattle (METRO), Contract No. CW/F2-82, Task 48.08. Prepared by BOAS Inc., Seattle.

1989a Botanical Analysis. In *Evaluation of the Burton Creek Rockshelter, Site 45-LE-266. Lewis County, Washington,* edited by C.J. Miss, pp.15-19. Submitted to Randle Ranger District, Gifford Pinchot National Forest, USDA Forest. Prepared by Northwest Archaeological Associates, Seattle.

1989b *The Botanical Assemblage of Layser Cave, Site 45LE223.* Submitted to Randle Ranger District,

Gifford Pinchot National Forest, USDA Forest Service, Randle. Manuscript on file, Botana Labs, Seattle.

1990   *The Flotation Results from 45LE220, 45LE223, and 45LE3410.* Submitted to Randle Ranger District, Pinchot National Forest, USDA Forest Service, Randle. Manuscript on file, Botana Labs, Seattle.

1993   Botanical Analysis. In *The Archaeology of Chester Morse Lake: Long-Term Human Utilization of the Foothills in the Washington Cascade Range,*edited by Stephan R. Samuels, pp. 12.1-12.2. Project Report No. 21. Center for Northwest Anthropology, Pullman, Washington.

1995a Botanical Analysis of 45KI428. Botanical Analysis of 45KI429. In *The Archaeology of West Point, Seattle, Washington* vol. 2, edited by L.L. Larson and D.E. Lewarch, pp. A7.1-A7.76. Submitted to CH2M Hill, Bellevue, Washington. Prepared for King County Department of Metropolitan Services (METRO). Prepared by Larson Anthropological/Archaeological Services, Seattle.

1996   Botanical Analysis of the Allentown Site, 45KI431; Botanical Analysis of Flotation Samples from White Lake, Sites 45KI438 and 45KI438A; and Botanical Analysis of Additional Botanical Samples from the Allentown Site, 45KI431. In *King County Department of Natural Resources Water Pollution Control Division Alki Transfer/CSO Site Allentown Site (45KI431) and White Lake Site (5KI438 and 45KI438A9 Data Recovery*, edited by L.L. Larson, pp. A3.20-A3.41. Submitted to HDR Engineering, Inc., Bellevue, Washington. Prepared by Larson Anthropological/Archaeological Services, Seattle.

Turner, N.

1975   *Food Plants of British Columbia Indians Part 1: Coastal Peoples.* Handbook No. 34. British Columbia Provincial Museum, Victoria.

1979   *Plants in British Columbia Indian Technology.* Handbook No. 38. British Columbia Provincial Museum, Victoria.

Turner, N., J. Thomas, B. Carlson, and R. Ogilvie

1983   *Ethnobotany of the Nitinaht Indians of Vancouver Island.* Occasional Papers No. 24. British Columbia Provincial Museum, Victoria.

13

Conclusions

Julie K. Stein

Objects dropped by people hold a fascination for all those who dream about the past. Archaeologists use a wide variety of objects and their relationships to each other (context) to test hypotheses about the past. This project brought together archaeologists, tribal members, students, and the public to excavate a shell midden threatened by erosion. Animals, metal, glass, stone, and charcoal recovered from Burton Acres Shell Midden are clues to the actions of the people who dropped them over the last one thousand years.

The excavation of the Burton Acres Shell Midden was conducted by the public with the blessing of the Puyallup Tribe of Indians for the purpose of salvaging the eroding site and educating citizens about the history of the First Peoples of Quartermaster Harbor. The site is endangered by bank erosion that will soon wipe away the remaining small portion of the site and spread it into Quartermaster Harbor. We had to act quickly to salvage the archaeological materials at the site, but recognized the site would be a good candidate for an educational opportunity. Over 700 people were able to experience first-hand the intricacies of uncovering the past by excavating four units. People learned that archaeology is more than just digging up objects, because they were invited to participate in the whole process: discovering, cleaning, and identifying the artifacts and their associations. The participants now understand the antiquity of the Puyallup people because they experienced the exhilarating process of piecing together many parts of the past, not just uncovering

isolated artifacts that are removed to a lab.

The challenge for the organizers of this project was to alter traditional archaeological methods to accommodate public participation. The solution seemed simple; each participant was to excavate one "bucket" of the site and follow it through the field process to the lab. They would screen, sort, and identify the contents of this bucket, and in this way realize that archaeology is not about digging for treasure. Rather they would learn that archaeology consists of a wide range of techniques, all focused on preserving the artifacts' contexts. Although the solution was simple, the logistics required inventing a new way of doing archaeology. Representatives of the Puyallup Tribe helped us with this process, incorporating a Native American voice throughout the experience (Figure 13.1). Students from Vashon High School and McMurray Middle School acted as "guinea pigs" so we could streamline the operation in the first days of the excavation. These students stayed with us and acted as tour guides to help onlookers appreciate the process.

Chapter opening photo: Volunteers, such as Annie and Devi Brule, enjoyed the whole experience. The project combined successfully the enthusiasm of the volunteers, the goals of the Puyallup tribal members, and the preservation of a cultural resource.

Figure 13.1 Many moments throughout the day were spent analyzing the project's various methods, as well as discussing the meaning of what was found. Holly Taylor (left) from King County Landmarks and Heritage Commission, Chester Satiacum (center) of the Puyallup Tribe of Indians, and Roxanne Thayer (right), a teacher at McMurray Middle School experience one of those moments, as they rest behind the check-in awning.

Most importantly, at the heart of the project, were experienced archaeologists who insured that the project maintained professional standards. They shared their expertise by teaching and encouraging each participant until the job was done. The archaeologists in this project were the helpers, the public learners were the doers, and the site was studied legally and professionally.

Unexpected to us, the people participating in this experiment not only appreciated and honored the subtleties of the process, but also embraced the task of interpretating the puzzle. Upon arriving, few participants realized that Native Americans lived throughout Puget Sound and in Quartermaster Harbor, often the very places inhabited today by the participants themselves. The 'archaeologists-for-the-day' used the artifacts they found to reconstruct the activities of the Native Americans who had used the site. Many participants were able to take the exercise one step farther by debating the accuracy of reconstructions based on artifacts alone. All understood why archaeologists concluded from the evidence that shellfishing and fishing were definitely important to the people who deposited the layers long ago, layers uncovered for the first time by each of these excavators. All realized that some of the fish species no longer live in the harbor and many of the shellfish can no longer be eaten, both having been affected by environmental degradation and

over-exploitation. They also realized that the people who created the Burton Acres Shell Midden were related to the Puyallup people with whom they were working, but who no longer live on the island. Everyone developed a respect for and genuine interest in the history of the Native American people whose material culture they were discovering (Figure 13.2).

THE PUYALLUP ANCESTORS

The written and oral histories, which tell us that the people known as the *S'Homamish* occupied Vashon Island and the surrounding area during the mid-19th century (pre-1856), made powerful impressions on the excavation participants. The *S'Homamish* were a signatory of the Medicine Creek Treaty of 1854. Even though this treaty gave these people reservation lands and the right to be enrolled at the Puyallup Reservation, some *S'Homamish* people did not move to the reservation land and remained in their home area for many years after the treaty was signed. The people who lived at the Burton Acres Shell Midden are the ancestors of the members of the Puyallup Tribe.

According to historical documents and oral histories, these ancestors were saltwater people, skillful sailors, who relied on the water for food and the raw materials needed in day-to-day life. Generally sharing in the cultural practices established among the Southern Coast Salish people, the *S'Homamish* joined others in seasonal food gathering camps or at temporary villages, and hosted friends and kin on special occasions. They enjoyed the abundant fish and shellfish resources from Puget Sound as well as the vast root and berry crops. Their diet also included land mammals and birds of many varieties. Multiple families lived in large cedar plank houses that contained built-in sleeping platforms, storage space, drying racks and fire hearths. The rich oral histories we have today indicate the importance of sharing stories that featured local landmarks and regional histories.

THE ARCHAEOLOGICAL DISCOVERIES
Public Involvement

In addition to the historical facts about the Native Americans that lived in this very spot, the participants were also surprised by the precision required in archaeological methods. Each volunteer was asked to fill out multiple forms, the contents of which were checked repeatedly during the process (Figure 13.3). Many people commented on the number (and color) of the forms and made comparisons to various government agencies! The importance of the forms was emphasized by the professional archaeologists guiding them throughout the process, which made an impression on the volunteers. In response, the volunteers filled out forms and recorded accurate provenience information seriously and carefully. The concept of "context" (the association of objects to each other and to natural features) was stressed, and the relationship between forms and the preservation of context was repeated often. People walked away from this experience with archaeology having a new appreciation for the importance of preservation, and for the irreversible damage caused by disturbances to a site.

People were also amazed by the amount of time required to sort and identify the remains from just one small bucket (Figure 13.4). Five hours were required to sort completely the content of a two liter-sized bucket of shell midden. Over the course of the project, thousands of small fish bones were separated from thousands of shellfish fragments. Charcoal was carefully picked out, as were flakes of rock, metal, and glass. The rounded rocks were to be returned to the site after they were weighed. The work required a systematic patience, an ability to sit in a hunched position, a good pair of tweezers, and sharp eyes. Only two people out of 700 failed the challenge, fleeing the lab area before their bucket was sorted. Others enjoyed the task. By participating in digging, completing paperwork, and sorting material, people gained an appreciation for the skills required to be an archaeologist.

Age of the Site

How old were the objects – when did people drop them here? The site contained layers full of artifacts that ranged in depth to 60cm below the surface of the ground. Relative and radiometric dates indicate that people lived and collected resources in Quartermaster Harbor over the last 1000 years. The deepest layers contained a few chipped-stone and ground-stone tools and lacked metal artifacts. The radiometric ages for these layers ranged from 1000 to 300 years ago. The layers found closer to the surface contained the same materials found in the deeper layers with the addition of a significant number of metal and glass artifacts. The radiometric ages for the shallower layers range from 200

Figure 13.2 Darlene Kenney bushes material in her screens as she asks questions about the process and the artifacts she sees.

years ago to the present. Lucy Gurand is recorded to have lived on this very spot around A.D. 1920s. She and her family, along with her ancestors from 1000 years ago, could have left the objects we uncovered.

Historic Artifacts

The majority of historic artifacts from Burton Acres Shell Midden point to conclusions consistent with written and oral histories, yet they give us details not recorded in documents. The objects made of metal, ceramic, and glass date from A.D. 1860 to the present. Given the influence of the Hudson's Bay Company within the region in the early part of the 19th century, the lack of older historic artifacts is surprising. The site may not have been occupied during the early 19th century, or perhaps was occupied only sporadically and for short periods of time. On the other hand, Native Americans may have lived at the site but simply did not use Euro-American trade items. Few personal items were found at the site. One way to interpret the lack of personal items at the site is that houses or other buildings where these objects would have been stored were not located in the excavated area. People kept their valuable personal items on their persons or at homes located somewhere else. They came to the Burton Acres area to dry fish and shellfish, bringing the tools required for these tasks. A limited number of square cut nails were found at the site, which could indicate that a small building or shed

Figure 13.3 R.J. Thomas, a student assisting in the excavation and member of the Muckleshoot Tribe, checks in volunteers and is first to emphasize the importance of filling out the forms carefully.

once stood there or that herring rakes (inserted with nail barbs) were used for harvesting herring from Puget Sound and Quartermaster Harbor.

Stone Tools

The analysis of stone tools and debitage (waste flakes produced during tool manufacture and use) suggests that chipped stone tools were not manufactured intensively at this site but were used there. Since only 79 chipped and ground stone artifacts were found (as compared with hundreds of other sites), little can be inferred about stone tool use at the site. Although the majority of the stone artifacts consist of debitage, nine retouched specimens were recovered, including two identical, unilaterally-barbed, triangular projectile points. These are unique to the region, and their function remains a mystery. The bulk of the lithics were made from "non-local" rock types, and as such, could have been part of a local trade network. This analysis reveals that tools made of stone were used throughout the past 1000 years, even during the past 200 years when metal tools would have been available.

Bone Tools

Although the volume of material excavated at this site is quite small, a significant number of bone and antler tools were recovered. Because tools made of bone and antler are both obtained from animals, they are often considered together. The bone and antler tools

found at the Burton Acres Shell Midden indicate that people worked wood and fished at this site. Three antler haft fragments and a chisel were found here, and are commonly tools associated with woodworking. A composite toggling harpoon valve and a bone point suggest activities associated with fishing. Approximately nine tools per cubic meter were found, which is more that four times the number of bone tools per cubic meter recovered from other sites in the Puget Sound region. This may be explained, in part, by the exceptional bone preservation at the site. Nearly all of the bone tools were recovered from the lower layers of Unit 29/58 where abundance of shell and the inundation of ground water from tidal fluctuation protected the bone. This high density of bone tools may also indicate a heavy-use or dumping area where tools accumulated in greater than average concentrations. Interestingly, almost all of the retouched lithics were also found in this part of the site. This association suggests that people worked nearby and dropped or tossed their broken tools and discarded fishbone and shells into this one area. We just happen to find the area of discard rather than the area where they worked (Figure 13.5).

Bird and Mammal Bones

The bird and mammal bones give us a clue about activities that may have taken place at this point of land. First, the numbers of bird and mammal bones found at the site are much lower than the number of fish, suggesting that fish were a more important resource at this particular location. Second, the types of mammals are the same from the earliest to the latest occupations of the site. Third, in Unit 29/58 there is an increase in the fragmentation (larger proportion of broken bones recovered from the smallest 1/8 inch screen relative to the larger screen sizes) in Layer 2C (post-contact), relative to the previous pre-contact layer (2D and 2E). Lastly, a greater percentage of bones are calcined (burned) in post-contact layers.

Many participants of the excavation developed interpretations about the bird and mammal bone. The small number of bones found suggested to them that these animals were not hunted or cooked at this spot. The change in the condition of the bones over time indicates that the people who first occupied the site were not breaking, crushing, and burning bones in the same way that later people did, because the bone found in the deepest layers was not as broken or burned compared to the bone in the shallower layers. When archaeologists find bones that are crushed and burned they interpret that bone as being processed in a fire. A favorite explanation of the participants is that people first dumped bone and shells (burned and broken) near the water, but over time the discarded shells are inadvertently burned and trampled leveling the area for more drying activity. This could very well have been the case. Most archaeologists suggest similar interpretations but often leave them out of their written reports, because they are difficult to test with the information we find. We include these ideas here as a hypothesis for future research.

Fish Bones

Tens of thousands of fish bones were recovered at the Burton Acres Shell Midden. Over 9000 were analyzed, providing us with an inventory of species collected. From oral histories we know that fish were an important resource to these people, and that some fish were more valuable than others. With this analysis, we can compare the historically valuable fish with the inventory of fish actually collected at the site.

The vast majority of the fish remains found in all stratigraphic units was Pacific herring. The Puyallup tribal members and local fishermen provided an explanation for the abundance of herring found in the site. Schools of herring traveled north into Quartermaster Harbor and turned into shore close to Burton Acres Shell Midden. The point of land on which the site accumulated extends into the harbor and would have been an excellent spot to capture the herring as they approached shore. Some Vashon Island residents who visited the excavation remember Puyallup people setting nets here and dipping for the herring from canoes. The point of land has the additional advantage of being a windy spot, catching the breeze whether coming from the north or south. Herring could be dried in this dependable wind, making the spot ideal not only as a place to catch the herring but also to dry them. The abundance of herring bone found in the site layers representing the time before and after Euro-American contact supports the observations preserved in the oral traditions.

Salmon remains are present in much smaller quantities throughout the site, which surprised the archaeologists. Salmon used to run into Quartermaster Harbor at high tide over the Portage Spit that connects

Figure 13.4 On this sunny day sorting is pleasant for a volunteer and archaeologist Laura Andrew (right). As they sit and identify items many questions are answered and hypotheses are generated.

Maury Island and Vashon Island, and we initially assumed that catching those salmon was the main reason for people living at the Burton Acres Shell Midden. In the early 20th century, Portage Spit was filled to make a road connecting Vashon Island to Maury Island, blocking salmon from that entrance to the harbor. Puyallup people describe watching the salmon swim over the barrier at high tide on their way to Judd Creek to spawn and remember fishing for them in Judd Creek. Because the entrance is now blocked and the creek's and harbor's waters are polluted, there are almost no salmon spawning in Judd Creek today. Our data suggests merely that the salmon caught here in the past were not processed at the Burton Acres Shell Midden.

Other taxa of fish found in the midden, which no longer live in large numbers in Quartermaster Harbor, are rockfish, flatfish, and sculpin. Even though they are found in small numbers at the site, the presence of these fish remains suggests that they used to be more abundant in the harbor and were important to the people living there at the time.

Shellfish

Shellfish, as expected, are the most abundant animal remains found in the Burton Acres Shell Midden. The site would hardly be called a shell midden unless shell was present in abundance. The richness reflects the importance of shellfish in the people's subsistence. Although a variety of taxa were recovered from the site, nearly 60% of the shellfish were Native

Figure 13.5 Kathy Duncan (left, facing camera) brought a group of students from the Jamestown S'Klallum tribe to participate in the excavation. The area where they dig, close to the eroding bank, seems to be a place where many objects were discarded rather than manufactured and used. Many broken tools and fragmented animal bones and shells were found.

Figure 13.6 At the end of the day on July 3, 1996, Angela Linse (top) and Julie Stein (right) excavated the deepest layers of the site before the excavation permit expires. The last volunteers had left as students and archaeologists carefully remove the last of the deep deposits in the unit.

Littleneck clams and Butter clams, both of which are found in the intertidal zone of the nearby shore.

The diversity of shellfish taxa remains fairly consistent through time from the pre-contact layers to the post-contact layers. The differences, however, are that the pre-contact layers have more large shell pieces as compared to the smaller shell pieces found in the upper layers. The pre-contact layers also have more robust species such as Horse clam, while the upper layers contain smaller size species such as barnacles. Additionally, more shellfish are burned in the upper layers than those in the lower layers.

Participants distinctly noticed the changes in sizes of shell pieces, because it took longer to sort the more fragmented shells in the upper layers than it did to sort the shells in the lower layers. The fragmentation of shell supported the interpretation that people first discarded shell, bird bone, charred plants, charcoal, and mammal bone, and later landscaped the area by spreading the shells. In this scenario, the landscape modification caused the shell to break into smaller pieces, which we in turn had to sort.

Charred Plants and Charcoal Remains

Lastly, excellent preservation of plant remains provides an opportunity to examine the economically useful plants at the site. More than 50% of the botanical assemblage is conifer. In particular, Douglas fir seems to

have been selected as an especially good fuel for fires at the site. Pine and Douglas fir bark, which also were found in significant quantities at the site, might have been used to steam shellfish and fish. Only about 25% of the assemblage is composed of hardwoods, which according to oral histories, are preferred material for shafts, foreshafts, prongs, hooks, drying racks, digging sticks, baskets, fish traps, bows, and arrows. Willow, which was also found, may have had a medicinal value. Elderberry, blackberry (or perhaps wild raspberry) and nuts may have been eaten at the site. As seasonal indicators, these berries suggest spring and summer occupation of the site.

CONCLUDING REMARKS

Only four units were excavated at Burton Acres Shell Midden, yet we now know some significant facts about the past. We know the age of the occupation and its durations across an important period of culture contact. We know that the encounter with Euro-Americans changed almost none of the activities that brought people to the landform; they continued to fish and collect shellfish. There was an addition, however, to the kinds of artifacts they used in those activities, most notably metal and glass.

Some objects were found that warrant special consideration. Two projectile points were found together in the last hours of the excavation. All participants had gone home and the archaeologists and some students excavated under lamplight to remove the deepest layer before our permit expired the next day (Figure 13.6). The points found side by side have identical shapes, each with a tang on one corner of the base that is longer than the tang on the other side. People couldn't help but notice the similarity between the shape of these points and the "Star Trek" symbol worn on the shirts of the actors in the television and movie series. Thus, we started to refer to these points as the "Star Trek" points. Similarly-shaped points have been found at English (British) Camp (45SJ24), and at Old Man House (45KP2), but to discover them here in Southern Puget Sound was surprising. These points would have been unusual at any site, but to find them in the last buckets, removed on the last day, in the deepest layer, made them something we will not forget.

Another remarkable discovery is the few fragments of eggshells recovered in the flotation samples. These are special not for their informative value (we imagine that bird eggs were part of people's diets), but rather for their remarkable preservation and the fact that we could recover such delicate artifacts. For some of us, capturing something so small and fragile gives us confidence that we retrieved as much as modern archaeological methods allow.

Everyone at the excavation had his or her favorite object or favorite moment associated with the site. For MaryAnn Emery it was the Washington National Guard button. For Judy Wright it was the dime that she found and the fact that so many people now know that the ancestors of the Puyallup were the first residents of Vashon-Maury Island. For Laura Phillips it was the adze handles, broken but preserved together in the position they were discarded; also, important to her is teaching people about the importance of properly caring for collections. For Sage Alderson-Gamble it was leading the tours and being part of the archaeology team. For Holly Taylor it was protecting the rest of this site and disseminating information about historic preservation. For me, it was the successful collaboration between the Native American community, the public, and the archaeologists, and the fact that we all learned from each other in ways we never expected.

The thousands of objects, bones, and rocks collected from this excavation are now stored in the Puyallup Tribal Headquarters. The entire collection is professionally curated and cared for according to state-of-the-art museum standards. We can all be proud that we participated in the acquisition of this endangered information and be relieved that it will exist, protected, for all to study.

Archaeology is piecing together a jigsaw puzzle when one has only a few pieces and no picture on the front of the box. The Burton Acres Shell Midden contributed many pieces to the puzzle of Puget Sound history, while kindling the imaginations of the people who observed and participated in the project. Education is a powerful preservation tool. When people understand the process behind archaeology and the importance of context, they willingly and knowingly protect cultural properties. Many people who participated in this project are now watching this site as well as their local landscapes for evidence of the past, and for ways to protect their pieces of the puzzle.